MIND WARS

MIND WARS

BRAIN RESEARCH AND NATIONAL DEFENSE

by Jonathan D. Moreno, Ph.D.

Dana Press

New York • Washington, D.C.

Published by Dana Press
New York/Washington, D.C.

DANA
PRESS

The Dana Foundation
745 Fifth Avenue, Suite 900
New York, NY 10151

900 15th Street NW
Washington, DC 20005

DANA is a federally registered trademark.

ISBN-13: 978-1-932594-16-4
ISBN-10: 1-932594-16-7

LIBRARY OF CONGRESS CATALOGING-IN-PUBLICATION DATA
    Moreno, Jonathan D.
        Mind wars : brain research and national defense / by Jonathan D. Moreno.
            p. cm.
        Includes bibliographical references and index.
        ISBN 1-932594-16-7
        1. Medicine, Military—Research.   2. Medicine, Experimental.
    3. Brain—Research.   4. National security.   5. Human experimentation in medicine.
    I. Title.
    UH399.5.M67 2006
    355´.07—dc22
                            2006006258

Text Design by Kachergis Book Design
Medical Illustrations by Kathryn Borne

www.dana.org

For Regina

A sister I bequeath you, whom no brother
Did ever love so dearly:
  —Shakespeare, *Antony and Cleopatra*

CONTENTS

ALSO BY JONATHAN D. MORENO

Is There an Ethicist in the House? On the Cutting Edge of Bioethics (2005)
Undue Risk: Secret State Experiments on Humans (1999, 2001)
Deciding Together: Bioethics and Moral Consensus (1995)

COAUTHORED BOOKS

Ethics in Clinical Practice (1994, 2000)
*Discourse in the Social Sciences: Strategies for Translating Models of Mental
    Illness* (1982)

EDITED WORKS

Ethical Guidelines for Innovative Surgery (coeditor) (2006)
In the Wake of Terror: Medicine and Morality in a Time of Crisis (2003, 2004)
Ethical and Regulatory Aspects of Clinical Research: Readings and Commentary
    (coeditor) (2003)
*Arguing Euthanasia: The Controversy over Mercy Killing, Assisted Suicide, and
    the "Right to Die"* (1995)
Paying the Doctor: Health Policy and Physician Reimbursement (1991)
The Qualitative-Quantitative Distinction in the Social Sciences (coeditor) (1989)

ACKNOWLEDGMENTS

Mind Wars grew out of a wide variety of conversations and experiences. Among those who provided me with specific assistance on problems I confronted as I explored this largely uncharted territory were Alta Charo, Missy Cummings, Tim Garson, Paul Gorman, Hank Greely, Jonathan Haidt, Gregg Herken, Dave Hudson, Steve Hyman, Daniel Langleben, Paul Lombardo, Clive Svensen, Jonathan Marks, Stephen Morse, LeRoy Walters, Cheryl Welsh, and Stephen Xenakis. Thanks as well to those who permitted themselves to be interviewed and quoted in these pages, as their participation helped fill in many blanks in my attempt to gain a comprehensive view of the topic. A number of others who preferred not to be identified helped me understand the culture of national security agencies in relation to science and academia.

I am particularly grateful to Floyd Bloom, who gave so generously of his time in reading the manuscript and corrected a number of errors in characterizing the relevant neuroscience.

The University of Virginia Medical School, one of the most collegial places in the universe, has provided me with a wonderful professional home. Much of the manuscript was prepared during a sabbatical at the Center for American Progress in the spring of 2005, where I was fortunate to have an intern, Jonas Singer, who holds an undergraduate degree in neuroscience. Jonas advised me on organization and assisted me in identifying sources and interpreting material for early drafts.

Over the years I have benefited more than I can express from attendance at science meetings as a bioethics adviser at the Howard Hughes Medical Institute; my exposure to some of the world's top neuroscientists through HHMI has been an extraordinary intellectual opportunity and has deeply informed this book.

The idea for *Mind Wars* began to germinate following an e-mail from the Dana Press' remarkable editor, Jane Nevins, who invited me to contribute to a special issue of the press' journal, *Cerebrum*. Out of that piece, entitled "DARPA on Your Mind," came an invitation to write a book-length treatment of the subject. I have since learned why Jane is so well regarded in the neuroscience world. Her persistent enthusiasm for the project carried me over many bumps in the road, especially when my confidence waned, and her observations and suggestions have never failed to be on target. The final product is a model of collaboration between author and editor.

Leslye, Jarrett, and Jillian have endured yet another book. As always, their patience and support mean the most.

MIND WARS

INTRODUCTION

ON HUMID HUDSON VALLEY SUMMER DAYS, the pale blue sky is domi-nated by enormous, billowing clouds that seem nearly close enough to touch. It was on one of those days, a Saturday morning in 1962 when I was ten years old, that a yellow school bus pulled up at my father's twenty-acre sanitarium a stone's throw from the river, disgorging about two doz-en young men and women. A distinguished psychiatrist, my father was the founder of techniques that have now been absorbed by our culture: psychodrama, role playing, group psychotherapy, and sociometry. I orga-nized an impromptu softball game with the group in a lovely field next to the building that served as a mental hospital and group therapy training center.

We played for about an hour, then it was time for them to get to work and I collected the bats and balls. I surmised they were either mental pa-tients or psychology students in for a training weekend, but at the time I gave little thought to which category they might fit. My father's pioneer-ing psychiatric treatment attracted all sorts of eccentrics. Yet it was un-usual for a group to show up in a school bus, so the incident lodged itself in my mind.

Years later, when I was a college student, I asked my mother, who worked closely with my father, about that weekend. "Oh," she said, "that was a group of patients referred to your father by a psychiatrist in Man-hattan. They were here to try LSD as part of their therapy. It didn't work."

LSD. Hmmm. . . . Why not, I asked.

"Well," she replied, "we couldn't tell where the effects of the drug end-ed and their symptoms began."

And did either of my parents try it along with their patients?

"No," she laughed, "but that young man on our staff, Harry, did."

And what about the school bus that had piqued my curiosity in the first place?

"Did they come in a school bus?" she asked. "I guess they rented it in the city. I didn't remember that."

Thus, what started as a minor mystery turned into a far more provocative tale.

I remember that until the mid-1960s, my father's office displayed a tax stamp signifying that as a physician he was authorized to order certain controlled substances from government-approved producers. The document read: "Cocaine, Marihuana [that was the exotic spelling one sometimes saw in those days], and LSD-25," which was the technical term for that formulation of the hallucinogen; somewhere along the line as LSD became part of pop culture, the "25" got dropped. Before that happened, federal officials reclassified the compound, recognizing that it was becoming popular as a recreational drug, so that it became much harder to obtain legally, even for doctors engaged in research. My father's tax stamp disappeared.

Between that weekend with the young patient group from New York and the time I casually asked my mother about the incident, LSD went from being an obscure chemical compound with some odd properties to an icon of the era. So did its chief proponent, Timothy Leary, who set up shop on an estate in nearby Millbrook, New York, scandalizing the locals. It was well known that he had been obliged to leave a promising career as a Harvard junior professor as he pursued his dream of personal and social revolution through psychedelically driven insight.

I recall my father's bemused attitude about the young Ivy League dropout psychologist who set up his LSD commune just down the river. Although at the time most assumed that Leary's interest in LSD must have made him a lone nut in Cambridge, insiders like my father, a frequent Harvard lecturer, knew better. In fact, Leary inherited a fascination about the drug from senior Harvard professors, a number of whom had been studying it since the early 1950s. The main difference was that Leary and some of his young peers took their interest to an extreme, seeing "acid" as a key to far-reaching social and psychic change.

Still more remarkable and known to even fewer was that much of the research on LSD and other hallucinogens was supported by national security agencies. In one of those great historic ironies, the hippie guru Leary and one of the central lifestyle aids of the 1960s youth movement of love and peace could trace their roots to America's early cold war defense establishment.

Thirty years later, my encounter with the weekend LSD-psychotherapy group connected to my professional life in a way I could never have predicted. In 1994–95, I took a leave from my job as a bioethics professor in a medical school to work for a presidential advisory committee investigating secret human radiation experiments sponsored by the U.S. government since the 1940s. My job was to retrace the often classified history of government support of human experiments. That's when I learned that LSD and other ways to influence the brain were of great interest to the CIA and the Pentagon until at least the end of the 1960s.

Still, it took the better part of another decade for me to achieve the insight that led to the idea behind this book: if national security agencies had so much interest in how the relatively primitive brain science of the 1950s and 1960s could help find ways to gain a national security edge, surely they must be at least as interested today, when neuroscience is perhaps the fastest growing scientific field, both in terms of numbers of scientists and knowledge being gained.

This hypothesis came to me only gradually, as a result of a number of associations and experiences. My book on the history and ethics of human experiments, *Undue Risk*, included some discussion of experiments with LSD and mescaline conducted by the Defense Department and the CIA, but after it was published I didn't see what the next step might be. Then I started to attend numerous seminars on neuroscience and began to appreciate what a burgeoning field it is. I also had the opportunity, through various professional connections, to meet and talk with many neuroscientists. They are some of the brightest and most creative people I have ever met and are working in an incredibly complex field. Next, in 2002, I was invited to speak at a national conference on neuroethics, organized by Stanford University and the University of California, San Francisco, a meeting that spawned intense academic interest in the ethics of neuroscience.

In the weeks and months after the conference, I began to take note of the frequent but casual references to national security agency funding in reports about amazing new neuroscience findings. I also noticed that many of the most prestigious neuroscientists I knew were being supported by some of these agencies' contracts. Yet when I raised questions about the specific nature of the national security interest in this work or the bigger picture behind it, the conversations tended not to go very far. Many of the scientists didn't know much about the larger context, didn't seem to have given it much thought, or figured it was an opportunity to fund their research that wouldn't lead to anything questionable.

When I had fully formulated the idea for this book, I turned to a number of individuals who were knowledgeable about the scientific world and its relation to government. They confirmed my hunch that the security establishment's interest and investment in neuroscience, neuropharmacology (the study of the influence of drugs on the nervous system), and related areas was extensive and growing. However, no one had attempted a systematic overview of developments in neuroscience as they might affect national security, nor had anyone raised the many fascinating ethical and policy issues that might emerge from this relationship. This was the case in spite of the fact that magazine and newspaper articles about some of these remarkable experiments often mentioned in passing that one national security agency or another was sponsoring the work. Rarely did the writers pursue the question of the agency's particular interest in the research or its role in the larger mission of the agency. I found it amazing that no one had attempted an analysis of the various pieces of brain science and technology in relation to national security and how they fit together.

I should note at this point that I am no loose cannon. I am deeply entrenched in the nonthreatening, even boring, academic establishment. I've taught at major research universities, hold an endowed chair at an institution not known as a hotbed of radicalism, am an elected member of the Institute of Medicine of the National Academies, have given invited testimony before both houses of Congress, and have served on numerous federal advisory committees. I have also been an adviser on biodefense to the Department of Homeland Security.

Thus, I felt comfortable taking on this controversial topic. But I encountered a level of sensitivity I had not anticipated. Since virtually nothing had been written about it, I decided to interview a few neuroscientists to get their sense of what national security agencies might have in mind for brain science, but I quickly ran into a wall. It turned out that the scientists who were working on intelligence agency contracts in particular weren't interested in talking for the record. In one case, the chief academic officer of a major university suggested I talk to one of his colleagues, but that professor declined and suggested instead that he connect me to his CIA contacts. They, too, weren't interested in being interviewed about the agency's interest in neuroscience.

The process of assembling the book wasn't damaged by this reluctance. Plenty of other neuroscientists and experts were willing to talk about the issues. Alumni and consultants of the Defense Advanced Research Projects Agency (DARPA), a science agency that is a crucial player in this field, were exceedingly helpful, though my efforts to obtain an interview with a current DARPA official about a report that the agency had its own ethicist failed. In addition, because DARPA mostly funds the nonclassified work of university scientists, an enormous amount of information is available on the public record.

Nonetheless, I found curious the reluctance to discuss the social and ethical issues concerning neuroscience and national security. After all, this was a time of enormous public discussion about the terrorist threat posed by biological and chemical weapons, when scientists and government officials were quoted daily in the popular press about the potential for an attack. For example, in summer 2005, the *Proceedings of the National Academy of Sciences* published a paper describing the operational potential of poisoning milk tankers with botulism, despite some serious misgivings on the part of the Department of Health and Human Services. And I had just been appointed to a committee to advise the government on biodefense analysis and countermeasures. Yet while biologists were writing and talking to the media about the potential for biodefense measures to go wrong, I couldn't persuade neuroscientists to talk for the record about the downside of their own field's involvement in national security work. What was going on?

I'm not the only one who has noticed the lack of ethical discussion among neuroscientists on the national security applications of their work. In 2003, just a year before I started to write this book, a debate about this very matter erupted in the normally collegial pages of the magazine *Nature,* one of the world's most prestigious science magazines. Publication in *Nature* is a key to tenure and promotion for university scientists, so when their editors are critical of an area of science you can be sure that community will notice. When the magazine questions the ethics of a whole field, something very unusual is going on.

In "Silence of the Neuroengineers," *Nature* editorialized that "the researchers [neuroscientists doing applied research] should perhaps spend more time pondering the intentions of the people who fund their work." Noting the amount of brain research funded by DARPA, the editorial observed that "the agency wants to create systems that could relay messages, such as images and sounds, between human brains and machines, or even from human to human. In the long term, military personnel could receive commands via electrodes implanted in their brains. They could also be wired directly into the equipment they control. Do neuroengineers support these goals?" the editorial pointedly asked. "Their research could make it happen, so they have a duty to discuss their opinions, and to answer questions from those who object to the development of such technologies."

Yet, the editors reported, many DARPA-funded scientists with whom their writers spoke "were reluctant to debate the potential military uses of the technology, saying that the agency's goal of brain-machine interface was still many years off." The *Nature* writers' experience mirrored my own.

The editorial brought sharp reactions. Three scientists from the California Institute of Technology observed that many military-funded technologies have brought positive results to society, and argued that each technology must be assessed on its own merits and not in light of the source of the funding that created it. A similar defense of the brain-machine interface research came from Alan Rudolph, a highly respected scientist in DARPA's Defense Science Office, who noted that millions of people with physical disabilities would benefit from new kinds of prosthetic devices. This kind of consideration should form the basis of ethical evaluation, Rudolph said, rather than the funding source. This heated exchange foreshadows the themes I will explore in this book.

I have concluded that several factors account for the lack of discussion described in the *Nature* editorial. First, scientists in general have an understandable reluctance to jeopardize relationships with research funding sources. Why should they talk to me when what they say could be perceived as embarrassing by those who are supporting their work? (Such misgivings are not unique to neuroscience, of course.) Second, at least some of the work that is being done requires a security clearance, and although I didn't want anyone to say anything that could be compromising, it is not always easy to remember or to be sure what is and is not sensitive information. There is an elevated level of concern about disclosing potentially harmful scientific information in this post-9/11 era. Again, this situation doesn't affect only brain scientists doing national security–related work. Other scientific fields are also involved in classified research.

But unlike the management of microbes or even nuclear fission, doing things to the brain gets into especially sensitive territory. Biologists mess around with microbes and physicists push electrons around, but people who work on the brain and examine ways to systematically affect it are getting really personal. Combine this perception with the fact that conspiracy theories about secret government experiments on the brain abound, and you have a recipe for extreme caution on the part of scientists who do brain studies and technicians who look for ways to influence the brain. Yet somehow, the obstacles to open ethical debate that the *Nature* writers and I encountered need to be breached.

Although a great deal of this book is about the possible national security implications that stem from high-tech neuroscience, such as various uses of neural imaging devices, not everything I discuss is neuroscience in the strictest sense. To give the historical context, I will reconstruct the federal government's long-standing interest in the behavioral sciences, support that proved crucial to its early development. I will also talk about a number of drugs and hormones that are of interest to national security agencies. Depending on the kinds of studies being done, these drugs and hormones might or might not be considered as building on the base of modern neuroscience, but they qualify as measures that might change brain chemistry or structure and hence the capacities of the subject. Similarly, I discuss devices that act on the peripheral nervous system and that target our ability to hear and smell. While they might not be breakthroughs that are directly related to neuroscience, they are of interest due to the way they can affect the brain and nervous system and provide some advantage in a conflict. So when I use the term "neuroscience" in this book, I mean it as shorthand for all the different ways that progress is being made in understanding and managing our mental processes through the study of the brain and nervous system.

Neuroscience, neuropharmacology, and various novel devices are being touted as presenting possibly significant advances in some areas of interest to national security agencies. Many, perhaps most, of these possibilities will not pan out or will lead in unexpected directions, as science generally does. But the very process of exploring them raises fascinating, difficult, and sometimes disturbing questions for social ethics and public policy. These prospective methods for learning about and influencing the brain are all fair game in an assessment of the prospects for the mind wars that surely lie ahead.

1:

DARPA ON YOUR MIND

The long-term Defense implications of finding ways to turn THOUGHTS INTO ACTS,

if it [sic] can be developed, are enormous: imagine U.S. warfighters that [sic] only need

use the power of their thoughts to do things at great distances. (Emphasis in original)

—*Strategic Plan, Defense Advanced Research Projects Agency, February 2003*

NEARLY EVERY WEEK, I take the bucolic drive between Charlottesville, Virginia, where I direct the University of Virginia's bioethics center, and my other home in Washington, D.C. On one of those drives a few years ago, I received a peculiar call on my cell phone. Like any loyal American, I pulled over before I answered.

"Dr. Moreno?" came a female voice on the line.

"Yes," I said.

"I need to talk to you about a matter, actually, it's . . . a national security matter."

"Uh, yes?"

"I read your book. I have been the victim of a government experiment and I need to talk to you."

As I had done many times, I explained to the caller that I was no longer working for the government on ethical issues about state-sponsored human experiments, that my top secret clearances had lapsed, and that I had long been back at my day job as a bioethics professor. I expressed sympathy about my inability to give her relief. Nonetheless, like others who have called or e-mailed me in the past eight years, she was sure I could somehow help her.

Mercifully, I lost the cell signal and the call.

JUST BECAUSE YOU'RE PARANOID DOESN'T MEAN SOMEONE ISN'T FOLLOWING YOU

I believe those who think they have been victimized by government mind control experiments are misguided. Yet, there are thousands of such persons. Many associate their ideas with conspiracy theories. I have worked for two presidential commissions and have been a member of several government advisory committees. During periods set aside for public comments in these meetings and in private conversations, I have heard many of these people provide seemingly lucid testimony about scenarios I find fantastic. Some of them are courageous and resolute in the struggle they perceive as having been foisted on them; others are distraught and terrified of what horrors the next day may bring.

Despite the vast distance between their worldview and mine, I have long been impressed at the irreducible kernel of truth behind their bizarre obsessions: that interest in understanding and manipulating the brain, while always strong, has flourished in recent years, particularly among those scientists in the United States and elsewhere who have been supported by the national security establishment. Often this interest is generally but misleadingly referred to as "mind control." The tale of research on the mind/brain is complex, rich, and rather odd; an offbeat slice of our social history.

Fascination with this idea that something like mind control is possible is by no means limited to the Western world. While I was lecturing in Pakistan in the spring of 2005, a senior psychiatrist told me that mind control by the CIA or other intelligence agencies is a common complaint of his patients. And just around the time I was in Karachi, the *India Daily* editorialized that "defense scientists and research engineers are busy all over the world in many countries trying to create the ultimate mind con-

trol machine that can make the enemy surrender without any fight," allegedly by manipulating the electromagnetic field around a person. Somewhat undermining the credibility of the newspaper's technology reporters
was the further remark that "those who had a close encounter with aliens
and extraterrestrial UFOs report that they communicate though their
mind and so speech."

A wide range of brain-related scientific endeavors, some as spectacular as mind control and others as mundane as political propaganda, has
also been pursued in the interest of the defense of the nation. Moreover,
the potential for emerging developments in the neurosciences and national security is indeed remarkable; old-fashioned notions of mind control are quite archaic compared with what is just over the horizon. The improvement of soldiers' war-fighting ability, brain-machine interfaces, and
the use of drugs and other measures to confuse and disrupt the enemy are
the sorts of approaches that are going to be developed over the next decades, driven by cutting-edge science. And that's not all. Recalling the epi-
graph at the top of this chapter, one might well wonder, for example, what
"things" the Defense Advanced Research Projects Agency (DARPA) has
in mind to do "at great distances." Later in this book, I will describe experiments in which a monkey has been trained to manipulate a computer
mouse or a telerobotic arm "simply by thinking about it." What else might
be made possible? How might such capabilities be applied?

THE FUTURE IS NOW

Doing things at a distance is also mentioned in a 2003 written statement for a congressional committee by DARPA Director Tony Tether. The
goal, he also said, is to exploit "the life sciences to make the individual
warfighter stronger, more alert, more endurant, and better able to heal."
DARPA's Continuous Assisted Performance (CAP) program, the statement continues, "is investigating ways to prevent fatigue and enable soldiers to stay awake, alert, and effective for up to seven days straight without suffering any deleterious mental or physical effects and without using
any of the current generation of stimulants."

These remarkable goals would be easier to dismiss if the agency did not
boast such an impressive track record. DARPA's overall mission is to bring

discoveries from fundamental research to bear on the requirements of to-day's warfighters, accelerating the pace of applicable discoveries. Among DARPA's accomplishments in its continuous effort to "fill the gap" between basic research and military use are the Saturn rocket, ground radar, the Stealth Fighter, and the Predator missile. DARPA-developed unmanned aerial vehicles have been used in Afghanistan and elsewhere. DARPA de-signed the computer mouse and, to give the mouse something to click for, the innovation that might prove to be the most socially transforming of them all: the Internet, first called the Darpanet. To be sure, about 90 per-cent of the agency's ideas fail, such as the one about a mechanical elephant intended to stalk Vietnamese jungles, but the ones that work are remark-able. As one high-ranking DARPA official put it, "DARPA is about trying to do those things, which are thought to be impossible, and finding ways to make them happen."

Such mechanical, electronic, and biotechnological innovations require extraordinary foresight, intelligence, and patience. Unlike in other areas of government, in the DARPA framework decades of development are ac-ceptable. Today, the agency is turning some of its considerable ingenu-ity to innovations in neuroscience. Early in 2006, DARPA announced its funding initiative for the coming fiscal year under the program "Applica-tions of Biology to Defense Applications." By my count, most of the agen-cy's desired research proposals directly or indirectly involve the brain:

- Biological approaches for maintaining the warfighter's perfor-mance, capabilities and medical survival in the face of harsh battle-field conditions;
- Biological approaches for minimizing the after-effects of battle inju-ries, including neurotrauma from penetrating and non-penetrating injuries as well as faster recuperation from battlefield injury and wounds;
- Approaches for maintaining the general health of deployed troops;
- Bio-inspired systems;
- Biomolecular motors and devices;
- Biological approaches to the growth of materials and devices;
- Understanding the human effects of non-lethal weapons;
- Micro/nano-scale technologies for non-invasive assessment of health (e.g., vital signs, blood chemistry);

- Technologies to enable remote interrogation and control of biological systems at the system/organ/tissue/cellular/molecular scales;
- Investigation of the interactions between physical forces, material and biology (e.g., interface of biology with magnetics);
- Novel mathematical and computational approaches to characterizing and simulating complex biological processes;
- New technologies to drastically reduce the logistics burden of medical treatment in the field;
- Advanced signal processing techniques for the decoding of neural signs in real time, specifically those associated with operationally relevant cognitive events, including target reduction, errors, and other decision-making processes;
- Novel interfaces and sensor designs for interacting with the central (cortical and subcortical structures) and peripheral nervous systems, with a particular emphasis on non-invasive and/or non-contact approaches;
- New approaches for understanding and predicting the behavior of individuals and groups, especially those that elucidate the neurobiological basis of behavior and decision making; and
- Technologies to engineer field medical therapies at the point of care, such as production of multiple drugs from a single pro-drug, or to adapt therapies for wide variations in body mass, metabolism, or physiologic stress.

The secret of DARPA's success is not its funding—at around $3 billion, its budget pales beside the research and development budgeting of U.S. spy agencies—but its brilliant use of intellectual capital. Its "*only* charter is radical innovation," according to its strategic plan. DARPA is a science agency, not an espionage outfit. (In fact, the agency historically has tried to stay away from spy projects.) It cycles top-notch scientific talent through its system just long enough that they don't get too jaded in their outlook. About 90 percent of DARPA's budget supports university research on vital human problems, including many basic medical studies. So although much of the science I will describe is DARPA-funded, and raises important policy questions, its largely open culture is generally praised by scientists as a smarter operation than the often more closed science programs of other national security agencies.

Since its founding in 1958 in response to the Soviets' Sputnik satellite, DARPA has been the key Defense Department agency whose mission is the pursuit of highly speculative scientific possibilities. Ironically, partly because so much of the agency's funding has gone toward science that doesn't seem to have an imminent national security payoff, the U.S. Congress has threatened to cut its budget in recent years. Though DARPA is only one national security agency among others that seek to exploit new technology, its relative transparency has made it a kind of symbol and an easy target for critics. No wonder the DARPA alumni scientists I spoke with said that the agency is especially publicity-shy for an outfit that does mainly unclassified work. Yet, to "sell" the Pentagon on a project, DARPA managers have to show that their idea fulfills some military need, however remote. So, what DARPA manages to get its Pentagon masters to fund tells something about what the military finds interesting. And, of course, not all DARPA projects are open access; the Stealth Fighter was one of the most closely guarded military secrets of the twentieth century.

It's hard to assess the efficacy of other taxpayer money spent on science. The official research and development budget for the Department of Defense is around $68 billion. And that figure doesn't include related national security research efforts supported by the Pentagon's secret, or "black," budget, which in the 1990s was often estimated in the press at about $30 billion but is surely higher now. Assuming that the proportion of R&D to operations in the secret budget is about the same as it is in the Pentagon budget, black R&D funds would be in the neighborhood of at least $6 billion. But these numbers are highly speculative and shouldn't be relied upon. At best, they would reflect only line items that can easily change, and what is included in the category of "research" is somewhat arbitrary. In addition to uncertainties about how much is being spent on research in general, we have no way of knowing whether the CIA itself is also working specifically on the potentialities of the brain sciences and various methods of enhancing or impairing human performance. What is clear is that DARPA is only one of several government agencies deeply interested in these and similar possibilities and that, other than for DARPA, citizens can't get much access to information about how their money is used.

Whatever the actual amounts at their disposal, the national security agencies spend a substantial portion of their resources on research support to some of our most brilliant scientists. In fact, our defense infrastructure is critically dependent on civilian talent, and vice versa. The results of this collaboration help advance basic scientific knowledge and can improve future health care, as well as address military and security questions. Someday, breakthroughs in the understanding and treatment of brain diseases such as Parkinson's and Alzheimer's might well be attributed to work paid for by DARPA. Techniques to study and augment the cognitive powers of healthy people, many examples of which I will describe later, might also prove effective in circumventing the destructive effects of these terrible diseases. As one Caltech professor told *Nature* magazine, "The military has always been visionary when funding neuroscience."

The onrush of discovery about the brain and the concomitant technological advancements suggests many areas of interest from the military or national security standpoint. Two of these, improving intellectual endurance and achieving mental control at a distance, are mentioned in DARPA's strategic plan. Others, such as memory enhancement and distant brain scanning (a device that could detect telltale blood flow in certain neural systems some distance from the subject), also hold interesting possibilities at the intersection of neuroscience and national security. Still others, such as a deeper understanding of the neural processes associated with stress and how to manage it, could affect the preparation of combat personnel as well as treating their stress reactions. This work not only offers to improve the human condition, it also presents formidable ethical questions that our society has barely articulated, let alone carefully addressed: How far should we go to enhance human performance, particularly our intellectual and emotional capacities? What adjustments in social systems will need to be made in light of these developments? Will such interventions have unintended consequences for societal institutions? What long-term risks are faced by those who are first to go down these paths?

These questions are especially pointed when we consider that the nature of human conflict could undergo basic change as the new neuroscience is applied to war planning. In a sense, all warfare ultimately happens between our ears. If opponents believe they have been defeated, then that becomes the reality, hence the military's investment in psychological operations such as propaganda leaflets and disinformation, despite their uncertain payoffs. But if targeted interventions are made possible by the greatly enhanced knowledge of the brain and nervous system now being generated at a feverish pace in our top neuroscience labs, complemented by ingenious new engineering and pharmacologic products, the battle of the brain will have truly begun. The powers that can claim the advantage and establish a "neurotechnology gap" between themselves and their adversaries will establish both tactical and strategic advantages that can render them dominant in the twenty-first century.

Besides remarkable enhancement possibilities that stem from new knowledge about the brain and new technologies developed in part for nonmilitary applications, national security agencies are engaged in research and development on drugs and devices that work through the senses to affect the nervous system. "Nonlethal weapons" include anesthetic agents, foul-smelling chemicals, and acoustic technologies that might be especially useful in civil disturbances and, in theory, morally superior to more violent measures that are out of proportion to the threat. But they are not so easy to control in the field and might run up against international treaties about chemical weapons.

It is ironic that discussions about national security often fail to include the optimal means of ensuring that people are safe to live their lives: keeping the peace. The sad fact is that there is a specific marketplace for the materiel of war, not of peace. On that front, it is important to learn what insights the neurosciences might give us into nonviolent means of settling disputes. Clearly, violence has been a constant theme of human history. But somehow, we have survived and, as a species, prospered. The neuroscientific studies now taking place in laboratories, as in the case of medications intended to act as "calmatives" in potentially violent situations, may point the way toward enhancing prospects of peaceful resolution. But the same work misapplied could diminish them.

WAR, ETHICS, AND THE BRAIN

"All's fair in love and war" must have been coined by someone who never suffered the worst excesses of either romantic or military affairs. The rules of war that have been introduced throughout history, from St. Augustine's just war theory to the Geneva Conventions, resulted from revulsion about what war can produce. Though cynics might dismiss the idea of rules in warfare, there are practical as well as principled reasons for some limits to what nations can do to each other in periods of armed conflict, especially to prevent our own soldiers or citizens from becoming the victims of barbaric acts if they are ever under the control of our enemies. With neuroscience now the object of national security agencies, it's time to consider whether new rules are required and, if so, whether these should be internal government policies, international standards developed by organizations like the World Medical Association, or, eventually, treaty obligations. Mental manipulation that is more insidious and perhaps more effective than torture and measures that are not biological or toxic have largely not been subject to explicit examination.

Neuroscience has undergone remarkable growth in recent years. One measure of this growth is the success of the Society for Neuroscience, which was founded in 1970 and now has over 35,000 members. Papers, books, and academic programs on neuroscience have undergone a similar explosion. It's become difficult to keep up with the new knowledge being created, let alone master the basics of all the disciplines that neuroscience tries to integrate: calculus, general biology, genetics, physiology, molecular biology, general chemistry, organic chemistry, biochemistry, physics, behavioral psychology, cognitive psychology, perceptual psychology, philosophy, computer theory, and research design. Sometimes called "brain science," neuroscience is really the science of the nervous system in all its glory and astonishing complexity; the nature and significance of the nerve fibers that run throughout the body have only recently been understood. Modern neuroscience was born from the integration of all these different fields of study and from new technologies for studying the brain in living people.

Several aspects of neuroscience raise novel ethical, social, and legal questions about the ability to monitor brain functions as they are happening and to associate them with psychological experiences, and the ability to use chemicals or other measures to change brain function. These abilities are double-edged, presenting tremendous possibilities for medical benefit and for misuse. Neuroimaging machines such as the functional magnetic resonance imaging (fMRI) scanner present the opportunity not only to study the way psychiatric disease is manifest in brain activity, but also to peer into how people think and to learn what they are thinking (at a very crude level so far).

There are indications that in time, neuroscience may be able to associate the subjective experiences of our inner life with objective events in our brains. Already, a lot of work is being done to correlate personality traits and responses in particular areas of the brain. The instantaneous recognition that an image of a face belongs to a racial group other than one's own has been correlated with activity in certain brain structures. Addicts' drug cravings are known to activate specific systems. Honesty and deception (and even self-deception) can be measured, and this capability will perhaps lead to brain-based lie detectors. Even thoughts about specific objects like cats and houses can be correlated with typical activation patterns.

Though drugs for restoring brain functions lost due to disease have been around for a while, their risks and side effects are still debated. The risk profiles of these medications are gradually becoming better understood, but, at the same time, other treatment approaches are emerging, such as magnetic, vagus nerve, and deep brain stimulation. Some researchers are beginning to explore whether these interventions may present hope for depressed patients. Closely related are brain-machine interaction projects to better understand the way the brain encodes and integrates data from its sensory, motor, and memory systems. These investigations might present options for treating people with paralysis, multiple sclerosis, and other motor impairments.

For example, transcranial magnetic stimulation (TMS) and drugs are being studied for possible improvements in cognition, with the focus on attention and memory enhancement. Greater basic science understanding of the brain and nervous system at the molecular level is feeding these developments. Drugs to treat attention deficits are progressing in their sophistication, including both medications for patients with pathological attention disorders as well as the more "normal" reasons for failure to concentrate, like fatigue, which might be managed with new pharmaceuticals for sleep regulation. Novel therapies for memory disorders are being explored with drugs that might improve the memories in people without a disorder as well.

As with any new medical intervention, enhancers affecting the brain will raise significant safety issues. Often, the effects of pharmaceutical agents are so subtle that they evade detection in short-term studies. Large numbers of individuals need to be exposed to them and followed for years to begin to ascertain unintended effects. A specific concern with memory-enhancing measures is that our ability to forget unnecessary information is an important evolutionary bequest. An avalanche of useless stored data could interfere with our ability to attend to important matters. We've never had to contend with this problem to the degree that neuroscience may soon make technically possible.

The brain is not like other organs; whenever we talk about modifying the brain, we enter special territory that calls into question whether we are stepping over some natural line and jeopardizing our essential nature or "personhood." Life presents obstacles and challenges. Is it wise to interfere with the struggle to meet these difficulties? Is it "cheating" to enhance the self through artificial means? Should we allow our personal identities to be medicalized and made the object of technological fixes? A deep and ongoing philosophical debate about enhancement technologies reaches back at least to Aldous Huxley's *Brave New World,* received impetus with the decoding of the DNA structure by Watson and Crick, and lies at the heart of the modern bioethics field.

Now, think about these philosophical questions in the context of national security issues. Everything I've described on the neuroscience research agenda, and more, applies in spades to the military context. If we can enhance the moods and cognitive capacities of soldiers, why would we hesitate? Or would we inadvertently weaken our forces by relying too much on external fixes? And is there something about the way wars are supposed to be fought or the way that national rivalries are to be pursued that precludes such "improvements?" I have already alluded to the many other interventions that may extend the range of action of individuals or reduce the chance of violence in resolving human conflict. Why should we shy away from them? These are issues I will return to throughout this book.

THE NATIONAL SECURITY STATE

The state's current interest in neuroscience is on a pathway that began decades ago. Understanding this pathway helps to establish the context for the critical issues that are raised in this book: What has been the role of government in science that may aid national defense? How should we assess the dependence of researchers on the support of national security agencies? How did the concept of "informed consent" arise from national defense–related experiments? Only after we comprehend the big picture of the relation of the modern state to science can we fully appreciate the moral and political issues associated with manipulation of the brain.

National security agencies haven't just discovered the importance of science in meeting the country's strategic objectives. At the outset of the cold war and now in the war on terrorism, exploiting the opportunities provided by the world's most potent scientific establishment has been high on the security agenda. In 1950, the National Security Council released a policy document called *NSC-68: United States Objectives and Programs for National Security.* The policy stated that "it is mandatory that in building up our strength, we enlarge upon our technical superiority by an accelerated exploitation of the scientific potential of the United States and our allies." More than a half century later, when President George W. Bush established the new U.S. National Security Strategy, he made a similar remark: "Innovation within the armed forces will rest on experimentation with new approaches to warfare, strengthening joint operations, exploiting U.S. intelligence advantages, and taking full advantage of science and technology."

As a result of this consistent commitment to science through all presidential administrations since World War II, the military-academic complex is now an integral element in the economies of our leading research universities. The Association of American Universities reported in 2002 that nearly 350 colleges and universities receive Pentagon research contracts, 60 percent of basic research funding. The leaders were the Massachusetts Institute of Technology, which was scheduled to receive half a billion dollars from DoD contracts in 2003, and Johns Hopkins, which was to get $300 million.

The federal government let science proceed on its own until World War II, when the Roosevelt administration saw the importance of this resource for the war effort. The most obvious example is the atomic bomb project, but in fact virtually every other area of science was also pressed into service by FDR and his advisers. This period set the pace for what followed in the cold war, when all sectors of society, especially science, were considered to be players in a sustained defense posture that had not been part of the American system before. To understand what political scientists and historians have called the "national security state" and the role of science in it, we need to go back to those postwar years.

The year 1947 marked a historic turn in the thinking of America's foreign policy establishment, a culmination of discussions that led to a new vision of the country's place in the world. Well before the end of World War II, it was clear that the United States would emerge as a dominant power, a position to which its economic trajectory had been leading since at least 1900, but one that the country had been unwilling to fully embrace in foreign affairs prior to the war. The isolationism that dominated prewar policy and that had evolved into virtually a national legacy since the administration of George Washington was no longer viewed as a viable guiding principle by the small group of major policymakers in President Truman's executive branch.

These men (they included George Kennan, Charles Bohlen, Vannevar Bush, James B. Conant, W. Averell Harriman, Robert A. Lovett, John J. McCloy, and James V. Forrestal, among others) were academicians, scientists, and corporate and financial leaders. They executed a rapid and vast expansion of federal powers, especially in the area of national defense. The goal was to confront an unprecedented threat in the form of the Soviet Union and to integrate science with military aims. That goal meant bringing the academic world's resources closer to the national defense system.

In fact, hardly a sector of American life was unaffected by the need to harness public energies to the task of protecting the nation in a dangerous new world. No longer could national defense rely on a mere "expedition," as was the case in World War I, and then retire to a demilitarized state on the old comfortable assumption that it was protected by a (largely) noninterference foreign policy and two oceans. Rather, the country would have to gird itself for a chronic war footing. Just as the Civil War transformed America's domestic character, so World War II transformed the face America showed to the rest of the world.

The United States was becoming a "garrison state," a phrase coined by Yale political scientist Harold Lasswell to describe a thoroughly militarized nation in a state of continuous conflict. The enthusiasm for preserving democracy abroad and confronting the Soviet Union on all fronts was dampened by concerns at home. Conservatives feared that the expansion of state power justified by a chronic wartime footing would threaten individual liberty and fundamentally alter the underlying values of the American experiment. Some leading liberals like Columbia's John Dewey were ambivalent about American involvement in World War II due to fears about the long-term consequences of militarization. These fears continued after the war in the form of congressional opposition to Truman's proposals to reorganize the national security establishment and change budget priories, initiatives that were ultimately successful. On the leading edge of American bolstering of military readiness and capabilities was the melding of military and scientific enterprise.

THE SCIENCE-SECURITY COMPLEX

Like other Enlightenment thinkers, Francis Bacon left behind a utopian vision of the future, a fictional state that expresses the author's philosophical aspirations for human society. In Bacon's version of utopia, *The New Atlantis* (1627), a scientific institute guides a society in a reasoned and prudent course that exploits the benefits of modern technology while steering clear of its dangers. To do so, Bacon explains, not all can be shared with the public: "We have consultations, which of the inventions and experiences which we have discovered shall be published, and which not; and take all an oath of secrecy for the concealing of those which we think fit to keep secret; though some of those we do reveal sometime to the State, and some not."

But with national security at stake, Bacon's idea that the scientific community would keep sensitive information out of the hands of government is now often turned on its head. Though a lot of published information is available about most government-contracted science (otherwise I couldn't have written this book), not all the results, applications, or analysis of security agency–funded research are made available to the public and the rest of the scientific community. There are good old-fashioned geopolitical and strategic reasons for this secrecy too. As the historian J. W. Grove has pointed out, the goal of government control over the "inventions and experiences" of science is not only national security, but also "commercial and economic dominance over other nations, and especially those nations that are considered potentially hostile."

The value of science in national defense was well understood by policymakers after World War II, when science played a critical role in the envisioned national security establishment. The First World War had already produced the National Research Council (NRC) to bring civilian science into military preparation and the war itself. Not a government agency, the NRC is today part of the National Academies, the first of which, the National Academy of Sciences, was chartered by the Lincoln administration to advise the federal government on science policy. The NRC continues to crank out many very influential reports, mainly at the request of Congress, that aim to articulate the policy implications of complex scientific, engineering, and medical problems.

The importance of the wartime alliance between science and the military for the World War II Allied victory is beyond dispute, not only in the most obvious examples of war-fighting technologies like sonar and improved armaments, and especially the atomic bomb itself, but also in less obvious ways through military medicine. A thick, two-volume report called *Advances in Military Medicine* published after the war attested to the many areas in which medical science enlisted for the war effort had achieved breakthroughs in basic understanding of human illness and in improved therapies, particularly those associated with the infectious diseases and bacterial disorders that plagued soldiers.

The most impressive single example of the benefits of government support of medical research during World War II was the rapid development of penicillin, the "wonder drug" whose refinement came too late to be of much use during the war but proved to be a great boon in the postwar era. Advances were also made in the treatment of malaria and other scourges. War is sadly but undeniably good for medical progress.

Academic scientists were largely untapped by government before the period leading up to World War II. Today, it is hard to believe that colleges and universities once feared and resisted the intrusion of the national government in their research priorities. This attitude changed drastically in the late 1930s as new academic leaders welcomed government support as a largely unexplored source of income but tried to minimize interference in scientific work, with mixed success. No institution exemplified this new cooperative spirit between government and the academy more than the Massachusetts Institute of Technology, which, under Karl Compton, devoted itself to the application of physical theory in solving a host of engineering challenges. Harvard President James B. Conant, a chemist by training, and MIT Provost Vannevar Bush, an early computer scientist who became president of Washington's Carnegie Institution, led the Roosevelt administration's crash program to realize the benefits of scientific research for the war.

Bush led the White House's Office of Scientific Research and Development (OSRD), which infused millions of dollars into scientific research during the war, mainly through the Committee on Medical Research (CMR). The CMR's projects included several that involved human experiments with conscientious objectors, prisoners, and persons with mental retardation. The OSRD was also a crucial player in the contentious wartime discussions about whether the nation's atomic energy resources would eventually be under civilian or military control or some combination.

Social scientists made important contributions to the war effort. The field of social psychology was of particular interest to military planners. An extended all-out war placed enormous social and psychological pressures on civilians. The Allies wanted to know how to turn psychological factors to their advantage to keep up morale at home and to dampen it abroad. Psychologists were brought in to develop sophisticated public opinion surveys, including ways of scaling attitudes. They produced studies to determine ways that the public's psychological state could affect economic activity and inflation, including absenteeism from work.

Valuable as the work on psychology turned out to be, it was far less dramatic than the atomic physics. The atomic bomb project showed that "big science" could be crucial to national survival and therefore that scientists were invaluable contributors to national security. After the war, the Pentagon established the policy that civilian science would become an integral part of military research and development, a process that was vividly described by Herbert N. Foerstel in his book *Secret Science*. After General Dwight D. Eisenhower announced this new policy in 1946, *Business Week* observed that "the odds are getting better all the time that pure scientific research will become, permanently, a branch of the military establishment." That same year, physicist Philip Morrison protested that a continued mobilization of science for war was "a dangerous and foolish state of affairs," and that "the armed forces are always sooner or later concerned with secrecy. . . . Such restrictions will greatly harm our science. It will become narrow, national, and secret."

Partly because of these worries, the new Atomic Energy Commission (AEC) was put under civilian control and, a few years later, the National Science Foundation was created to ensure a protected sector of civilian science. Nonetheless, funding for science within the military also grew, and with this support it wasn't hard for scientists, always in search of new funding sources, to be kept on the military hook. They were given research contracts that enabled them to pursue their most cherished projects, but on the condition that they also produce new knowledge of military value that would be kept classified. Their sensitive role was highlighted during the McCarthy era, when the FBI monitored the loyalty of scientists more closely than that of any other group, and of course the passing of so-called atomic secrets became the basis of the trial and execution of Julius and Ethel Rosenberg.

Yet science and secrecy are almost antithetical concepts. For science to advance efficiently, there must be wide dissemination of results. Science is not tinkering. It cannot be conducted by lone inventors in a garage, but requires an extended community of highly skilled and specialized experts in discrete areas. Attempts to practice science covertly are often self-defeating. Secrecy can also be used as a cover-up for incompetently executed science and as a means of avoiding public embarrassment, both for scientists and government officials.

Even if an attempt is made to render the communal process of science consistent with secrecy, it's often very hard to distinguish between information that's sensitive and information that isn't, and sometimes data that doesn't seem important for security purposes at the time gets published, only to be followed by attempts to withdraw it from the public years later. That's what happened after the October 2001 anthrax attacks, when the U.S. government decided to reclassify material that had long been publicly accessible, including on Internet Web sites. Senior physicians and scientists who are military veterans have told me of being admonished by nonscientist superiors to keep certain information quiet, in spite of the fact that it was already well known in the medical community. The process for manufacture of the atomic bomb is the classic example of science conducted in secret: the most important and highly classified scientific secret in history stayed secret only about four years, until the Soviets exploded their own device in 1949. For all the imagined and actual espionage activity around the bomb, competent physicists only had to study the published literature to get the main ideas.

As the cold war ended, two generations of American scientists had enjoyed the opportunities that came with military funding. For many researchers, access to government funding had become a way of life that even the National Science Foundation and the growing private sector could not completely replace. The argument for continuing this relationship was now economic rather than explicitly military competition. In 1992, the National Academy of Sciences recommended that DARPA expand to develop technologies that were of "dual use," valuable for both civilian commerce and national security. We will see that dual use is pervasive in national security agency funding of neuroscience research.

INFORMED CONSENT AND NATIONAL SECURITY EXPERIMENTS

During the immediate post–World War II period and in the early cold war, the long and intimate connection between science and the national security state precipitated the emergence of the idea that human experimentation demanded the subjects' "informed consent." Almost as soon as the new U.S. Atomic Energy Commission went into business in early 1947, it was confronted with the simmering conflict between military and civilian concerns about the use of people in national security experiments.

Late in 1946, the Manhattan Project's deputy medical director proposed that a report about secret plutonium injections be declassified. These injections had taken place at several major research hospitals in 1945 to learn about human excretion rates of plutonium for the sake of radiation worker safety. The scientific considerations that favored sharing data were met with a skeptical reception by an AEC declassification officer, who wrote on February 28, 1947, that this particular report, among others being reviewed,

> appears to be the most dangerous since it describes experiments performed on human subjects, including the actual injection of the metal plutonium into

the human body. . . . Unless, of course, the legal aspects were covered by the necessary documents [perhaps a reference to patient consent forms], the experimenters and the employing agencies, including the U.S., have been laid open to a devastating lawsuit which would, through its attendant publicity, have far-reaching results. [In addition,] the coldly scientific manner in which the results are tabulated and discussed would have a very poor effect upon the public.

A few weeks later, on March 19, the AEC's medical chief signaled his agreement with this point of view and the documents were kept hidden. Thus, the outcome of this early contest between scientific openness and the secrecy needs of national security was determined by the need for the emerging national security state to be spared the public embarrassment that would come with lawsuits and scandal from a shocking human experimentation.

As the implications of this incident became apparent, the AEC's nonphysician officials sought to protect the government from the risk of subsequent embarrassment about human experiments by requiring what they called "informed consent" of any subjects of future agency-sponsored radiation experiments, the first known use of this expression. However, some specific provisions of their proposal proved unacceptable to the AEC's medical scientists, particularly a requirement that the subjects sign a form signaling their willingness to volunteer. Following some tense sessions in which the AEC backed off from its original position, the agency's general manager remarked wryly, "Indeed, from the discussion at the meetings of April 3–5, it seems evident to me that doctors would not allow their judgment on this matter to be influenced by anyone." The state's requirements for self-protection were one thing, the medical profession's autonomy quite another. Surely, doctors could be trusted to do the right thing.

Various interests competed for attention in this episode: the state's interest in avoiding liability and damaging publicity and medical scientists' interest in protecting their professional autonomy and authority. Recall, too, the prewar concerns among academic leaders that the universities stood to lose their independence in setting the rules and goals of their research mission once they accepted government largesse. The resistance of medical scientists to the AEC's proposed policy can be seen as a manifestation of these federal contractors' determination to resume their roles as civilian university professors and insist on the self-direction of the academy, free of government control, even though it was government that was to provide both the funding and the radioisotopes for their research.

There was an event besides the plutonium injections themselves that worried AEC officials and motivated them to advance informed consent as a protection of state interests. The officials were aware of the trial of Nazi doctors taking place at Nuremberg in U.S.-occupied Germany at exactly the time they were dealing with the plutonium experiments declassification issue. Later in the 1940s, when the AEC and the Pentagon were jointly considering the development of nuclear-powered aircraft, questions were raised about how to determine that the power source would not endanger the crew. Human experiments were desirable, but the most likely population, long-term prisoners, invited disconcerting comparisons to the Nazi crimes. Indeed, evidence about the heinous nature of the concentration camp "studies" that was presented at the trial of the Nazi doctors made such an impression on the three American judges that they decided to write their own code of ethics for human experiments, which has come to be known as the Nuremberg Code.

The immediate influence on American medicine of the Nuremberg Code was, to put it mildly, minimal. A code of ethics written in response to Nazi crimes was easily dismissed as irrelevant to normal medical research. But government officials were sensitive to both its legal and political implications. The code surfaced in military medical planning as a direct result of President Truman's creation of a single Department of Defense from the Departments of the Navy and War. Undertaken in 1949, this massive bureaucratic overhaul predictably left numerous gaps in the new department's policy framework, including the field of human experimentation. There were fears that the Soviets were outstripping U.S. military scientists in the development of unconventional weapons and in the defense against them, including the conduct of human experiments. From 1950 to 1953, several internal Pentagon advisory committees deliberated on the matter of suitable guidance. In the end, both the military and medical members of all these panels largely agreed that a written policy was a bad idea, preferring instead to rely on an unwritten code of ethics and the virtues of those in charge rather than opening the matter up to legal scrutiny.

Once again, the traditional independence of medical scientists was cited.

At first, like the AEC leadership, the Pentagon's civilian leadership was reluctant to assert itself in what was turning out to be a controversial area. Truman's secretary of defense, Robert Lovett, had his people working the problem on a track parallel to the military medical advisory process. The job of finding an acceptable policy was handed over to Defense Department legal counsel, which identified the Nuremberg Code (incorrectly) as prevailing international law, but the code never was part of international treaty obligations. A key player in the decision to adopt the code was Lovett's deputy, Anna Rosenberg, who as an assistant defense secretary in charge of manpower, was the highest-ranking woman in the history of America's defense establishment to that time. She took the proposed use of the Nuremberg principles a step further by adding a written consent requirement, a consequence perhaps of her experience as a leading labor-management specialist in New York.

In spite of Rosenberg's advocacy and Lovett's sympathy for the proposal, internal opposition to any written policy at all was so strong that the civilians in Truman's Defense Department decided it would be best to refer the matter to the incoming Eisenhower administration, as the new Defense officials would have to live with the consequences of imposing an unpopular policy on the military medical bureaucracy. The matter was still urgent, the outgoing officials explained to their successors, because there was still no departmental rule that would allow important human experiments on the effects of atomic, biological, and chemical weapons. There is reason to believe that Lovett himself briefed the new defense secretary, Charles E. Wilson, on the issue. Lovett believed that the 1949 reorganization of the defense establishment needed more work, so he was involved in the transition to an exceptional degree. So it was that Wilson finally signed off on a top secret memorandum to Army, Navy, and Air Force secretaries on February 26, 1953, making the Nuremberg Code the Pentagon's policy, including Rosenberg's idea of written consent.

Secretary Wilson had been chief executive officer of General Electric, and at the time he took office, he served as CEO of General Motors. Like Anna Rosenberg, he was experienced in labor negotiations and approached the human experiments issue in a matter-of-fact manner as a contract problem, a deal between the department and the experimenter on the one hand and the human volunteer on the other. This would have been consistent with the management style he and Eisenhower agreed he should bring to the job, running the Department of Defense like an industrial corporation. Wilson also intended to implement Eisenhower's "New Look" defense strategy, important in its greater reliance on nuclear weapons. Any help the medical experts could give on human factors in atomic warfare would clearly be welcome.

The Wilson policy was forward-looking. Unlike academia, where there was virtually no discussion of human experiment rules in the early 1950s, the Pentagon technically required written voluntary consent for human experiments. Unfortunately, it was also highly implausible that a policy pronouncement would change the medical or military cultures of the day, neither of which was friendly to the idea of individual self-determination. Many would argue that even today informed consent is often honored more in the breach than the reality of doctor-patient relations, and self-determining soldiers don't have much of a future in the military. During the 1950s, the novelty of the idea that national security human experiments required a signed consent form posed an obstacle to implementation that not even official Pentagon policy could overcome. In 1975, following the revelations of LSD experiments in the military during the 1960s, the Army's own inspector general concluded that the policy was inconsistently applied at best.

NATIONAL INSECURITY

The national security state gains its power from the way it engages the resources of society as a whole, including science. Security is to a great extent a psychological rather than a political condition. It is instructive to recall Daniel Yergin's 1977 observations about the concept of national security as it functioned when he was writing during the cold war:

> We must remember that "national security" is not a given, not a fact, but a perception, a state of mind. . . .
>
> The doctrine is characterized by expansiveness, a tendency to push the subjective boundaries of security outward to more and more areas, to encompass more and more geography and more problems. It demands that the country assume a posture of military preparedness; the whole nation must be

on permanent alert. There was new emphasis on technology and armed force. Consequent institutional changes occurred. All this leads to a paradox: the growth of American power did not lead to a greater sense of assuredness, but rather to an enlargement of the range of perceived threats that must urgently be confronted.

There is an open-ended quality about the need for national security that should give us pause. Complete security is, of course, an ideal that can never be fully realized, but since national security is so important, it's easy to justify making science a partner in the quest. Politically, it's hard to oppose both national security and science (the real-world analog of motherhood and apple pie), so at some point the price tag becomes almost irrelevant. When a project is dual use and might advance science as well as security, the justification seems still greater. In the case of neuroscience research, which might well lead to cures or at least better treatment for some terrible diseases, critics of defense agency support make the Grinch look like Santa.

Just as national security sets up an open-ended goal, a similar point may be made about the "war on terror." As former Nebraska Senator Bob Kerrey observed while serving on the 9/11 Commission, terror is a tactic. "Terror-ism" is an ideology that advocates the use of terror to advance a political agenda. As long as humans are constituted as they are, this tactic will always have a target, so the prospects for eliminating either terror or terrorism from the human experience are not great. This situation contrasts with the less psychologically driven challenges of the cold war, which implicitly required "only" the collapse of the Soviet Union. As William D. Casebeer and James A. Russell have written, "What all students of terrorism realize is that the phenomenon if anything remains diffuse, not easy to define and in general difficult to understand." In its very ambiguity lies much of terrorism's political power.

It is possible to live with a constant terrorist threat, as the British and the Israelis have shown, without succumbing to it or to one's adversary. The more we learn about the ways our brain and nervous system deal with stress, the better we might be able to manage it. Advancing knowledge in the neurosciences can help us deal with terror, but even as the science progresses, there are plenty of other issues along the way: Do some of the results of neuroscience threaten our identities as individuals, as human beings? Can we use the tools that might be produced without risking our national values? Will the new resources the military will enjoy ultimately make us less secure? Who is going to have the final say about how these new assets are used or if they are used at all?

The relationship between science and the national security state in the context of a war on terror is still unfolding. Unlike the post–World War II era, when scientists who had eagerly joined the war effort saw military-related funding as a continuation of their previous employment, today significant distance lies between much of the scientific establishment and defense organizations. First, science has many other funding sources, including venture capital, that were not important players in the 1950s. Second, cultural differences between scientists and military officials bring with them a degree of mutual skepticism, if not outright suspicion, that was not the case fifty years ago, before Vietnam and Watergate. Third, unlike the experience of physics with the atomic and hydrogen bomb projects, the life sciences have not had much experience with operating under highly classified conditions. Many important researchers and their institutions chafe under security constraints, including not only sequestering their data but also tightening rules on the handling of pathogens in their labs and limiting visas for graduate students from abroad.

Since the fall of 2001, government funding for science has gradually been repositioned to address terrorism-related issues. The National Institutes of Health was the first off the mark among government research agencies, with programs to develop new vaccines and other treatments for exposure to some of the most feared potential biological weapons. Eight regional biodefense centers have been built since 2003 with $350 million in NIH grants. On the other hand, the new, bulky Department of Homeland Security was much slower to get its funding out to scientists, and labs funded by the Department of Energy that have a history of defense research are only gradually and painfully being retooled.

Still, if history is any guide, the war on terror will present great opportunities for renewing the traditional security-science complex. Perhaps the most dramatic Bush administration initiative is Project BioShield, which provided an initial outlay of $5.6 billion for new treatments for high-profile diseases such as smallpox and anthrax that might be used as weapons. A lobbyist for companies that stand to benefit from BioShield told me he considers this only the down payment, with billions more to come from Congress. Congress in early 2006 was considering more incentives for private companies to enter the largely unprofitable field of vaccine development, including generous patent protections and protections from lawsuits for injuries from new countermeasures to bioweapons.

The BioShield initiative dovetails with a new Food and Drug Administration exception to the standard requirements for permitting new human medicines to go on the market. Part of the 2002 Bioterrorism Act allows the FDA for the first time to approve certain drugs without testing them for effectiveness in humans if they are important and it wouldn't be ethical to do human experiments. Say, for example, that there's an experimental anthrax medication. Because it wouldn't be ethical to expose human beings to anthrax to test it, under the new rule the drug could be approved if it works in two animal species and has been found safe in people. Called the "two animal rule" by health policy wonks, this exception to standard FDA requirements could be a windfall to companies with drugs that might be important in a public health emergency, especially one suspected of being caused by terrorists, because they could avoid spending many millions of dollars on lengthy clinical trials.

One catch is that the only plausible market for these medications would be the federal government. Companies that go to the trouble of developing a product for a single potential customer take a significant risk, as do their investors. Only a few will have sufficient assets to take such risks. That's where the BioShield legislation comes in, to avoid one disincentive for getting into this area by providing public funds so the companies able to work in these fields don't incur crushing debts.

It's far too early to tell how the neurosciences will finally be affected by new worries about national security in an environment so different from that of the cold war, and how this burgeoning area of science will affect the course of the war on terrorism and our society. As I will describe in the next chapter, history can be a source of instruction about how far the ripples caused by the alliance of science and national security can spread.

2:

OF MACHINES AND MEN

❖

NEUROSCIENCE GIVES US THE OPPORTUNITY to study and manipulate living brains, but ancient anatomists learned what they could by cutting up and inspecting the brains of the dead. Physicians contributed some functional information by observing people with unusual diseases or injuries. Since all these people's work on neuroanatomy was conducted over the course of so many centuries, there are lots of different ways of dividing up the brain, which is a marvelously complex organ with lots of overlapping systems. Weighing only about three pounds but containing 100 billion nerve cells, or neurons, with more possible connections than stars in the universe, the adult human brain is evolution's greatest achievement.

YOUR FRIEND, THE BRAIN

The major anatomical divisions of the brain are the forebrain, the midbrain, and the hindbrain. The forebrain includes two major structures, the telencephalon and the diencephalon. The telencephalon can be divided into five parts, the most obvious of which are the left and right hemispheres of the cerebral cortex, which each contain two lobes. The sulci are deep grooves or folds that give the outer layer of the brain its wrinkled appearance. Within the telencephalon are four organs that are important in regulating emotion, memory, and some movements: the amygdala, the

THE FOREBRAIN AND BRAIN STEM

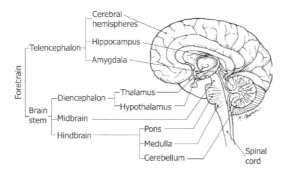

THE CEREBRAL CORTEX

THE BASAL GANGLIA

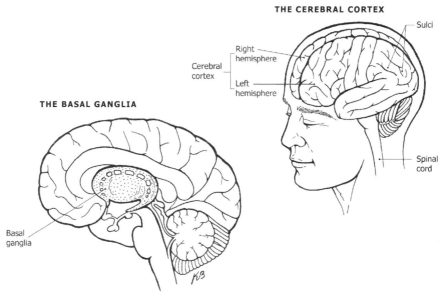

hippocampus, the basal ganglia, and the septum. The diencephalon is divided into the thalamus and the hypothalamus. The thalamus relays messages going into and out of the forebrain, while the hypothalamus relays information about the internal regulation of the body.

The colliculi in the midbrain transmit sensory information from the sense organs to the brain. In the hindbrain are organs that control the heart and respiration: the pons, the medulla, and the cerebellum; the latter stores basic learned responses and, apparently, information about the position of the body. The hindbrain, midbrain, and diencephalon together make up the brain stem and contain the connections between the cerebral cortex and the spinal cord.

At the microscopic level, the 100 billion neurons that make up brain tissue carry electrochemical messages throughout the system. They are the oldest and in some cases the longest cells in the body, extending from the brain to the spinal cord. There are sensory neurons, motor neurons, and interneurons (mostly in the central nervous system), which send information between the sensory and motor neurons.

This quick tour hardly does justice to the organ system that is increasingly the object of national security attention. Throughout this book I will cover a lot more territory about how much is being learned about what neural networks are associated with what activities, leading to a vastly enriched understanding of the brain itself. But what will all this tell us about the "mind?"

THE NEURONAL SYMPHONY

A dream of brain science is correlating neural activity with subjective intentionality. It has long been known that each neuron gives off a detectable electrical signal even though it's insulated. But monitoring with electrodes all the billions of neurons to find those that might be involved in a particular thought or act is impractical. What has not been appreciated until recently is that even simple actions require the activity of hundreds of millions of neurons, and the same neurons could be involved in many different acts. The science writer Carl Zimmer has likened this brain behavior to a symphony, an "orchestra of neurons scattered across the brain." A popular model of the brain known as connectionism predicts that the activity of the brain is often widely distributed rather than localized.

As a result, it isn't necessary to monitor the whole brain to find only those neurons that are involved in a certain act or thought, but one can tune into a subset of neurons to get information about many different actions. Using Zimmer's orchestral analogy again, you might only need to listen to a few instruments to figure out which symphony is being played. That's essentially the strategy that has allowed brain-machine interface experiments with animals to succeed: a small number of neurons are monitored, maybe only a few dozen, while the animal is trained to press a bar to get a drink. The neurons' activity is recorded and after the lever is disconnected from the drink, the animal learns that the only way to get the reward is by thinking about pressing the bar.

What's happening here seems to be more than just short-cutting the series of neurons needed to initiate some activity in the world beyond the brain. The instantaneous nature of neural-state reading that modern imaging technologies make possible indicates a mediated access into the thoughts themselves. The neuronal symphony being read out is the music of mind, access to which could be key to creating symbiotic relationships between human beings and their cybercreations. These relationships could include environmental arrangements or extend to the physical conjunction of humans and machines. The term "cybernetic organisms," or "cyborgs," was coined by two NASA scientists in 1960, an agency that had good reason to wonder how to complement weak human bodies with mechanical devices as it later aimed to meet President Kennedy's challenge of a lunar landing by 1970.

In 1985, I was invited to write a paper for a journal of speech pathology on the ethics of cochlear implants. Since my mother had to adapt to deafness in one ear as a young woman, any ethical problems with this new technology that could restore hearing in so many, albeit at substantial cost, were not clear to me. I began to appreciate the dilemma when I came to understand that those who had helped forge a culture of deafness had struggled to overcome the emotional cost of viewing oneself as disabled in a society that so values physical wholeness. They rejected the notion that those who were hearing impaired required an expensive mechanical gizmo to make them complete. At a deeper level, some critics suspected that a certain way of life was at hazard, or that the cochlear implant was a kind of electronic stalking horse for a barely perceptible set of future choices about the marriage of minds with machines, choices that are going to be made possible first in the realm of national security research.

TOWARD "WEARABLE ROBOTICS"

In the mid-1980s, I was on the staff of the Hastings Center, a bioethics research institute in New York's Hudson Valley. One week, a few of us

went to the Rehabilitation Institute of Chicago to help develop an ethics program there and hear about some of the ethical issues the institute
faced. Many of the patients at the institute were amputees, learning how
to function with primitive prostheses. Though my think tank colleagues
didn't know it at the time, I was very familiar with the problems encountered by amputees. When I was five years old, my mother, already dealing
with hearing loss, had her right arm and shoulder amputated after she was
diagnosed with a chondrosarcoma, a cancer of the bone marrow that is
rare in young women. The best anyone could have done in the late 1950s
was to outfit her with an artificial arm that hung at her side. She declined
and essentially rehabilitated herself, learning to write, drive, type, and sew
with one arm. (To top it off, she had been right-handed!)

My mother's situation has been difficult in many ways, but doesn't
compare with those most impaired by neurological disorders. At the re-
hab institute, a nurse talked about a patient she worked with who had
"locked in" syndrome. Following a massive stroke in the main artery to
the brain, he was unable to move except for the muscles that control eye
movement. Yet his brain worked just fine. He communicated by fixing his
eyes on particular letters on an alphabet board until he built a sentence.
As the nurse told us what it was like to care for this patient and how close
their relationship was, she confessed that there was one question she never asked him: Do you want to live? "After all," she said, "what would I have
done if he had said 'no'?"

Traditionally, patients who have neurological disease from strokes,
spinal cord injury, or head trauma are given nursing care and are left on
their own to recover what brain function they can. After several months,
they get back pretty much all they will ever have and need to learn how to
live with the results. But new therapies, including bionic limbs, computer
chip implants, and electrical stimulation of neurons, have the potential to
vastly improve the lives of people with these disorders, even giving hope
to those who have lost nearly everything and are locked in. These same
medical treatments, under the heading of neural engineering, also present
intriguing possibilities for military planners.

In particular, the progress in prosthetics is remarkable and heartening.
A German company called Otto Bock HealthCare has created the C-Leg,
one of a new generation of high-tech prosthetics that takes advantage of
new lightweight materials, computer sensors, tiny microprocessors, and

hydraulics. Some patients are so proud of their shiny robotic limbs that they are loath to conceal them under their clothing. As a 2005 *New York Times* article observed, both major universities and the U.S. military "are exploring ways in which people can be enhanced by strapping themselves into wearable robotics, or exoskeletons." Complementing normal human physical capacities such as strength and agility with the new prosthetics is already within reach, but is only the beginning of this story.

A HELPING HAND

DARPA's neuromics program, which is aimed at finding ways to permit brains and machines to interact, is a great example of dual use technology. According to DARPA's description, "the Human Assisted Neural Devices Program"—the acronym HAND is especially apt for one of the experiments I'll describe—"represents a major DSO [Defense Sciences Office] thrust area that will comprise a multidisciplinary, multipronged approach with far reaching impact. The program will create new technologies for augmenting human performance through the ability to non-invasively access codes in the brain in real time and integrate them into peripheral device or system operations." DARPA has so far invested $24 million in the brain-machine interaction program.

Here's a science fiction scenario: an army of robots capable of movement nearly as precise as that of a human soldier, each controlled by an individual hundreds or even thousands of miles away. These automata could undertake actions that would be foolhardy for human beings but worth the tactical risk for machines; because they are controlled by people, they would have the benefit of creativity that might limit even the most advanced android. But the old-fashioned remote control scenario would have the operator pushing buttons or moving levers while seeing on a monitor what the robot is seeing, a method that would be far too clumsy for the instantaneous reactions often required in combat. What is wanted is a technology that would allow the robot to respond as soon as the distant operator does. Such a technology would, in effect, have to be able to read the intentions of the operator, his or her thoughts themselves, not merely respond to the operator's muscle movements through a mechanical apparatus.

Some of the technical requirements for the soldier–extender robot army are, literally, within reach. It is already technically possible to transmit people's ideas about movement from the premotor cortex (so called because it's in front of the motor cortex) to a mechanical device. A Duke University neurobiologist has implanted contacts in a monkey's motor cortex and connected them to a computer that controls a robotic arm. After the monkey's own arms were tied down, an orange slice was placed at the end of the robotic arm near the monkey's mouth. The monkey's motor neurons fired while he tried to reach for the orange, moving the arm via the computer. Gradually, the monkey learned how to control the arm with greater precision. Then he realized he didn't even need to try to move his own arm to get the robot arm to move, but could make the device work just by thinking about reaching for the orange. His own arms became mere appendages.

This experiment and another that evolved from it get even closer to showing the potentiality of a brain-machine meld. The key to this other one appears to lie in the parietal cortex, which is located on top of the brain between the frontal and occipital lobes and behind the central sulcus. The parietal lobe or lobes, as there are two connected parts on each side of the brain, have two pertinent regions. One involves sensation and perception, enabling us to perceive or "cognize" the world around us. The other parietal region integrates sensory input to build a sort of grid that allows us to represent the environing world.

A group of Caltech scientists showed that intention can be read directly from activity in the parietal cortex. Wires were inserted into monkeys' parietal cortex, and before they actually touched a cursor to get a reward, their neural activity in "planning" to touch the cursor was recorded. The rewards were then varied and recordings of the specific expectations for each reward were made, translated from neural activation. The team was actually able to predict what reward the monkeys expected to receive based on the data they had on what neurons fired in their brains. In this case the researchers were not detecting the neurons related to muscle movement but the cells that correspond to *planning* the movement. According to some definitions, they were able to read the monkeys' minds.

Related DARPA-funded work at the University of California seeks to substitute damaged brain regions with an implanted microchip that will relay messages to their ultimate destination without normal mediation. The imprinting on the chip is based on the circuitry of slices from the rat hippocampus. The most optimistic forecast is that work like this might someday result in silicon-based treatments for neurological disorders like Alzheimer's and stroke.

This and several other DARPA-funded projects to create "neural prostheses" are intended to help neuroscientists understand the brain and how minds and machines can interact. Ultimately, decades from now, human abilities could be augmented so that combat soldiers could have vastly more powerful and faster robotic arms and legs, and pilots could control vehicles through intentional thought alone. Warfighters, intelligence officers, medics, and rescuers could wirelessly manage legions of robots through direct communication between the human brains and on-board artificial brains. Sensory feedback systems from the robot to the human operator would provide crucial information about the hostile environment. Even now, the first generation of what will be many different kinds of human-augmenters is crawling around university campuses: an agile six-legged computerized robot named RHex ("rex"). RHex is being tested with hand controls and computer programs, but if the monkey-prosthetic experiments bear fruit, within a few years RHex could be controlled via a brain-machine interface, and then the possibilities are immense.

DUAL USE AND BEYOND

The neuroscientists doing the work for DARPA obviously hope that these techniques can be applied to a technology that enables instantaneous control of devices for paralyzed patients and for those with amputations. Since the information is being gathered directly from the neurons involved in intention, the remaining challenge is an engineering task: to construct machines that are precise enough to do justice to the specificity of the ideas read off the brain's impulses. A company called Cyberkinetics has been given FDA approval to implant chips in the motor cortex of quadriplegics so that they can control a computer mouse. From that simple beginning, more refined mechanisms can be developed. One problem, as a company official pointed out, is that the brain-reading program may be too sensitive, that it may relay ideas about movement that are only being entertained as fantasies but not meant to be acted on. Presumably, avoiding embarrassing mistakes would take some practice.

Short-term implanted electrodes are already showing promise for some patients with neurologic disorders for whom medication and conventional surgery have failed. A few people with severe epilepsy have had their brains scanned to determine what areas were misfiring during a seizure, then electrodes were wired in to zap the malfunctioning region. The episode was prevented and patients reported that they hadn't even known when the minute electrical impulse had fired. In theory virtually any brain disorder could be amenable to neuromodulation—migraines, sleep apnea, obsessive-compulsive disorder, depression—if the troubled system is identified with sufficient precision. Thousands of people are wearing internal neurostimulators for epilepsy and otherwise intractable pain with promising results, but the batteries need to be replaced every five to ten years.

Practical problems with more complex long-term implants abound. Electrical leads are one thing, but there's a lot of work to be done with biomaterials before the implants could reside in a brain indefinitely. It's too dangerous to put them into a healthy human now. The early neurostimulators have been associated with several deaths, and infection rates are high. There are also no implantable devices that can keep track of the activity of individual brain cells with sufficient precision to achieve the kinds of cyborg scenarios that DARPA is aiming for.

Another problem is that, to be truly interactive, the robot arms and other brain-linked machines are going to have to give feedback to the brain. The brain-machine interaction concept must be a two-way street, not just an electrode zapping some tissue or a mind pushing a machine around. So, Duke researchers are also developing "tactile feedback." According to an article in *Technology Review,* transducers in the robot arm will identify when the arm has made contact with another surface and send electrical signals back to the brain, stimulating sensory regions and simulating touch. As Duke professor Miguel Nicolelis told the magazine, "The trick is to give the right kind of feedback so the monkey's brain will incorporate the robot as if it were a part of its own body." The key to full integration of the brain-machine interface might lie in nanotechnology, to provide what Nicolelis has called "direct links between neuronal tissues and machines."

In the same 2003 *Technology Review* article, a top NIH neuroscientist in charge of the Neural Prosthesis Program noted the importance of national security funding in these efforts. William Heetderks said that "DARPA is putting much larger resources into the area than has ever been seen before." He concludes that the money DARPA can provide "will have a tremendous effect." This dual use strategy has helped create a receptive climate for military-funded research, but it also raises the question whether the admirable goal of aiding those suffering with disabilities is tainted by DARPA's historic association with goals that will not necessarily excite such universal approval.

NO BULL: OF ROBORATS AND MEN

A variation on this concept of direct neuronal control of external devices is the external control of intentional activity. There was a death-defying experiment in a Spanish bullring by Yale's José Delgado in 1964. Delgado implanted a set of radio-controlled electrodes in a bull's brain, a system that he called a "stimoceiver." He stood at one end of the ring as the bull charged toward him, then pushed a button on a transmitter. The bull stopped short, turned around and calmly trotted away. The event deservedly got a lot of press attention, including a rave review in the *New York Times*, and I can recall seeing the amazing film on television as a child. A few noted that the work was Pentagon-funded, prompting fears that people could be turned into robots.

What is most striking is the fact that this line of work continues with Defense Department sponsorship to this day. Even more astonishing and doing Delgado one better is the "roborat," in essence a remote-controlled rodent developed through DARPA funding as part of its Defense Sciences Office Brain Interface Program. Scientists at Downstate Medical Center in Brooklyn (where I once worked as a bioethics professor) implanted electrodes in a rat's brain and ordered the rat to walk or climb through any path it was instructed to follow—navigating a maze, climbing ladders, going down ramps, climbing trees—through electrical impulses sent directly to certain brain centers. The roborat's operator used a laptop computer to control what would be better described as a living robot than a trained rat. This is how Dr. Michael Goldblatt, the director of DARPA's Defense Sciences Office, described the results to a San Diego science and technology group in 2002:

You'll see how this guy [the roborat] is going up, down, through a complex of mazes, and really he is being driven the way your children drive remote-controlled cars. He's driven right, left, and given a reward. It's nothing more than a typical kind of training that you always knew you could do. It's just cute and clever. Since then, we have outfitted this guy with a television backpack, which allows us to have a rat's-eye view of the environment that he's in. Here's the guy going through rubble for search and rescue as well. Then we've eliminated that backpack, and now we're using much larger rats. We can bury all of the electronics in the abdominal cavity and hide the electrodes underneath the skin so that in essence he looks like any garden-variety rat that you would find in a cave. He can be a very interesting tool. Since then, we've gone way beyond this. We've taken him out to field trials; we can go up fences, around fences, climb trees. We can go anywhere a rat would go, and a lot of places they wouldn't normally go. We've driven them past the Dunkin' Donuts and everything else, and we can keep them on track without any difficulty.

This project got widespread press coverage when it was published in *Nature* in 2002, mainly as a curiosity. Films of the brown-and-white rats with little powerpacks and receivers strapped on their backs like cartoon rodents going to school, scuttling down sidewalks, around barriers and up hills, made great television. Somewhat lost was the important scientific question of whether these kinds of experiments actually have the potential for enabling people with serious brain injuries to manage substitutes for their limbs. Not all the neuroscientists I have spoken with think that they do.

Whether or not the roborat and its brethren may advance serious medical purposes, there are other reasons for DARPA's interest: the possibility that remote-controlled robot animals could be used in dangerous situations like bomb detection and mine-clearing operations. As one scientist put it, "This is an animal with 200 million years of evolution behind it. Rats have native intelligence, which is a lot better than artificial intelligence."

The roborat's creators were quick to point out that they didn't subject the animals to such dangers, nor did their technique shock or cause pain to the creatures. Quite the contrary: the roborat's pleasure center was stimulated by one of the electrodes, while two other electrodes stimulated neurons associated with the left and right whiskers, so the rat could be "ordered" to go forward and left or right. If anything, the rat would have experienced the commands as pleasurable, because the satisfaction centers (for food, water, or warmth) were stimulated. All of the federal rules for humane treatment of lab animals were scrupulously followed.

Animal rights advocates might complain that the roborat experiment revealed that ethics rules could be strictly followed even in the course of turning an animal into a virtual robot. Although the concept of self-determination for rats excites little sympathy among humans, there are deeper issues here about appropriate limits of human control over other creatures, as well as potential forms of control over human beings. The first challenge is to establish exactly what the ethical problem is with direct control over the brains of others through neuroengineering. After all, human beings have trained animals for thousands of years, have domesticated them for tens of thousands of years, and have eaten them for millennia. Does the biblically endorsed human domain over nature and its creatures extend to roborats? What exactly is the difference between training a favored pet not to pee on the carpet with positive reinforcement and ordering him to pee on the fire hydrant by manipulating a switch on a panel that sets off an electrical charge that causes him to lift his leg?

Long-range control of simple animal systems doesn't need to be limited to electrical stimulation. In 2005, two Yale University neuroscientists reported that they were able to stimulate specific behaviors in genetically modified fruit flies. The flies were injected with the nucleotide ATP and then exposed to laser light that released the ATP and activated ion channels. Most of the animals jumped and flapped their wings following a pulse of light. This experiment is obviously highly contrived and very far from natural conditions, but it does enable scientists to study flies' behavior-control circuits and establish the principle that neurons can be remotely stimulated in all sorts of ways.

Although fruit flies and rats are not in any interesting way self-determining creatures, some believe that such direct neuroengineered control over another creature, no matter how modest or even repulsive, does threaten to turn our manipulable but still independent fellow creatures into mere appendages of human will. At least traditional animal training involves a sort of interspecies negotiation. With these robotic creatures, there is no such negotiation; they are simply divested of intent or purpose that is distinct from that of a human agent. The "otherness" of these distinct beings, however inferior they might be to humans on the evolutionary or cognitive scale, is virtually extinguished. To some, this will smack of a thoroughly arrogant attitude toward nature, a hubris that takes us down a slippery slope toward the domination and finally the destruction of our environment and ourselves.

But wait, these are rats we are talking about, at least in the DARPA-sponsored studies. Considering that countless human beings stand to benefit from this research, philosophical worries about human arrogance seem pretty abstract. So another line of criticism focuses on the fact that this is "dual use" research, that it has both a medical goal and a military purpose. With their acute sense of smell, roborats or other similarly controlled animals could be used in mine-clearing operations, or to scuttle into tiny holes in search of earthquake victims in collapsed buildings. And there shouldn't be any guesswork needed when the critters have found a survivor. University of Florida researchers found that rats produce a unique neural pattern when they find a scent they are searching for, confirmed so far in the odor of explosives like TNT and RDX, another boon to securing an area after a terrorist attack.

The prospect of an army of remote-controlled robots is chilling to some, but not all. Efforts to reduce the loss of life among one's own forces are surely fair game, and machines have been used in warfare for a long time, starting with catapults. But if the "I, Robot" image isn't a compelling objection to the long-term possibilities presented by this brain-machine research program, critics point out that the roborat technology could be used to control human beings as well as animals. "The use of animals in warfare is ugly enough without the further insult to their dignity involved in turning them into involuntary cyborgs," in the words of commentator James Meek in the *Guardian*. "And a military command committed to the use of creatures which are part-animal, part-machine, is going to be that bit less reluctant to interfere in its soldiers in similar ways."

Less likely but also technically possible would be "robohumans," putting aside for a moment the question whether they would provide any advantage over machines or well-trained self-directed soldiers who have high tolerance for risk. Already, an Associated Press reporter has written about his experience being electrically stimulated through a special headset while an operator moved a joystick from side to side. The electrical currents threw him off balance when he tried to walk forward, and he "felt a mysterious, irresistible urge to start walking to the right whenever the researcher turned the switch to the right." The researchers at Nippon Telegraph & Telephone Corporation claim that with galvanic vestibular stimulation, they can direct an individual to walk along a pretzel-like path. The company's benign goal is to enhance the learning of ballet dancers and others who undertake complex physical skills. "It's as though an invisible hand were reaching inside your brain," the reporter concluded.

ARTIFICIAL INTELLIGENCE AND INTELLIGENCE AUGMENTATION

DARPA's Information Technology Processing Office is pursuing projects intended to produce the ultimate artificial "cognitive information processing system," one that can reason on its own, learn from its experience and take direction, explain itself and reflect on its abilities, and respond to novel situations. Why would the Pentagon be interested in an artificial system that can think for itself? The actual sales pitch that DARPA threw its bosses to fund this field is unknown, but it is interesting to speculate. Consider a scenario that is familiar to science fiction buffs from *Dr. Strangelove* to *Terminator III*: the uber-computer program that manages all the offensive and defensive missile systems without the need for unreliable human involvement. Unfortunately, these programs always run aground due to their lack of spontaneity, their inability to adapt adequately to novelty. They are also often designed to resist human input. Hence, instead of steady and rational managers, they become doomsday machines. The ideal artificial cognitive system that is the goal of the DARPA effort would eliminate the annoying tendency of these systems to destroy the planet. Of course, given that the DARPA folks are the ultimate wonks, it's also entirely possible that they are really interested in the intellectual challenge such a project presents, with countless possible but highly speculative spinoffs (consider all those dumb machines in our lives that we wish would just reprogram themselves), and managed to talk the higher-ups into funding it.

In a sense, the perfected cognitive information system would be the ultimate result of classical computing, following along in the tradition of the genius Alan Turing in the 1930s, who defined the concept of computability. His "Turing machine" would be able to describe its own operations. Turing would have recognized DARPA's summary: "Given their abilities to process knowledge and to reflect on their own behavior, cognitive systems might be best characterized as *systems that know what they are doing*" (emphasis in original).

This DARPA effort is motivated by the need to anticipate the significance of the vast expansion in computer power to be expected in the twenty-first century (according to Moore's Law, doubling every eighteen months), with capacity likely to equal that of a primate brain within the next few decades. But how to manage all this power the way that the brain does? The current concept of a computer processor is just too slow to keep up. The problem is to provide an integrative system for truly intelligent computation, a qualitative rather than merely quantitative leap. Hence, according to a DARPA pamphlet for project proposers, some of the questions these projects should address:

- Given the vast amount of raw information that computers sort through almost instantaneously, can the human and animal perceptual systems give us insights into how to find important low-frequency events in huge amounts of data?

- How might a cognitive system learn the salient things from each experience it has and later use what was learned in an appropriate way to interpret and successfully cope with new situations? How can it find the right remembered experiences to apply to each new situation?

- How can we build systems that effectively keep an eye on themselves? How can an artificial reflective system operate in real time? Is there virtue in allowing reflective processes direct access to internal structures and processes, i.e., true introspection?

- What insights from neuroscience can provide breakthroughs in the building of artificial cognitive systems?

- Can insights from neuroscience and elsewhere inspire mechanisms to allow people to cope with the increasing problem of information overload?

Notice that the idea of a cognitive information processing system uses the primate brain as a model. It's within the Turing framework. But the remarkable experiments in brain-machine interaction invite a theoretical approach that goes beyond artificial models of cognitive information processing. I have described experiments that involve various permutations of organic-inorganic relationships: the brains of rodents and primates, machines like synthetic arms or laptop computers, and messages sent to brains from machines or brains sending messages to machines. But ultimately what is sought is a genuine loop of information between the two entities, true interaction that will lead to a symbiotic relationship between brain and machine.

The conceptual basis for this ultimate goal has its origins in the work of two pioneering thinkers and their creations, Alan Turing and his universal machine and Vannevar Bush and his Memex. Turing's major contributions took place just before World War II, leading to artificial intelligence as we know it today. Bush was provost at MIT and FDR's science adviser; publication of his Memex, part of his "intelligence augmentation" program, came as the war was winding down. The men were contrasting personalities. Turing was an abstract, difficult, and distant genius; Bush, a Washington power broker, perhaps the first scientist to wield great influence within a presidential administration.

In a paper in the journal *Semiotica,* the philosopher Peter Skagestad elegantly explained how artificial intelligence (AI) and intelligence augmentation (IA) have come to be complementary approaches:

> Both the Turing machine and the Memex attempt to mechanize specific functions of the human mind. What Turing tried to mechanize was computation and, more generally, any reasoning process that can be represented by an algorithm; what Bush tried to mechanize were the associative processes through which the human memory works. But the two machines also represent very different approaches to mechanization. Specifically, the Turing machine is a *digital* machine while the Memex is an *analog* machine. Digital machines work through discrete states and can represent continuous processes only in functional terms, i.e. in input-output terms: if, given the same input, the machine produces the same output as some natural process, the machine will be said to simulate that process, irrespective of whether the internal processing inside the machine in any way resembles that of the natural process.
>
> Analog machines, by contrast, utilize internal processes that resemble the natural processes they are simulating. . . . The Turing machine is a simulator while the Memex is a replicator. . . . The Memex, which attempts to replicate

human memory, and hence may be said to embody "artificial memory," was not intended to rival the human mind but to extend the reach of the mind by making records more quickly available and by making the most helpful records available when needed.

The result, as Skagestad explains it, was "two complementary inventions of networks and personal computers."

Together, the Turing idea of machines that think and the Bush idea of machines that think with the human mind constitute the key conceptual combination in the brain-machine work now being done. There is nonetheless a tension between them: Is the computer a mimic of mental operations (Turing), or is it an extender of the mind so that it can access information otherwise inaccessible to it (Bush)? In fact, though we are more familiar these days with Turing's digital paradigm than Bush's analog model, Skagestad's analysis suggests reasons that Bush may be on the ascendant. Citing the great nineteenth-century American philosopher and mathematician Charles Peirce, Skagestad notes that reasoning is not simply algorithmic but experimental. To extend intellectual activity we cannot rely on a device that is reducible to the mechanical semiotics of os and 1s, but must develop machines that can execute theoretic deductions through the semiotics of meaning that is an essential characteristic of cognition. As Skagestad points out, in a typically witty, provocative, and at first curious passage, Peirce wrote:

> A psychologist cuts out a lobe of my brain (*nihil animale a me alienum puto*) and then, when I find I cannot express myself, he says, 'You see, your faculty of language was localized in that lobe.' No doubt it was; and so, if he had filched my inkstand, I should not have been able to continue my discussion until I had got another. Yea, the very thoughts would not come to me. So my faculty of discussion is equally localized in my inkstand.

Peirce's point is not that his inkstand is equivalent to his brain, but that his ability to access new thoughts is intimately bound up with his ability to use the symbols that flow from his pen. These are the associations that are familiar in what Peirce's friend, financial benefactor, and sometimes intellectual rival, William James, called the stream of consciousness. External objects like pens, word processors, and symbolic notation are required both to propel us through the stream to new thoughts and to sustain the previous thoughts in passages too lengthy for us to entertain at once. These items augment our intelligence, as Skagestad puts it.

Will Charles Peirce, the philosopher of intelligence augmentation, be vindicated? Will the key to the future of the expansion of mental powers be Bush's analog device? In an era in which all things digital seem to have conquered the world, exactly that surprising turn may be on the horizon as the key to both bringing the reach of machine intelligence within our mental grasp and extending that grasp.

AUGCOG

DARPA's Augmented Cognition, or AugCog, program seems to operate on the philosophy that digital intelligence alone cannot achieve the full potential of functional relationships between minds and machines; analog processes also are required. Realizing that potential involves working toward a symbiotic relationship between computers and human beings. A paper by two human factors engineers, Dylan D. Schmorrow and Amy A. Kruse, describes the strategy: cognitive states are measured and tracked and the resulting data are used to augment the operator's environment and adapt it to the operator's condition. The result should be better performance in stressful situations with fewer people involved.

DARPA described AugCog this way in 2002:

> The *Augmented Cognition* (AugCog) program will extend, by an order-of-magnitude or more, the information management capacity of the "human-computer" combination by developing and demonstrating enhancements to human cognitive ability in diverse and stressful operational environments. Specifically, this program will develop the technologies needed to measure and track a subject's cognitive state in real-time. Military operators are often placed in complex human-machine interactive environments that fail when a stressful situation is encountered. The technologies under development in AugCog have the potential to enhance operational capability, support reduction in the numbers of persons required to perform current functions, and improve human performance in stressful environments.

In a project called "Augmented Cognition for Cockpit Design," scientists are working on improving several elements of mental activity: working memory for multitasking, executive function enhancements to remember information in context, increased capacity for sensory input, and attention to different priorities of a task. Each of these aspects of cognition is challenged by a particular system that can test improvements in the brain-machine interface: a computer-based command environment (working memory), an unmanned combat vehicle (executive function), a combat vehicle (sensory input), and an integrated individual combat system (attention).

Again, we don't know the exact functional augmentations that might result, but we can theorize about some desirable aims. Help with working memory might keep track of details in a complex environment that seem insignificant but might need to be called upon later. A drone combat vehicle could have the adaptability to execute appropriate actions in new situations, while a manned vehicle could augment and manage the natural sensory inputs of the operator with, say, auditory or visual signals beyond the human range. An integrated individual combat system would filter important elements of the fighting environment from less important ones. Over the long haul, the goal is to lessen the human workload in each area and for any combination of them. Similarly, in an AugCog project at the Cognitive Ergonomics Research Facility at San Diego State University, scientists are laying the groundwork for improving the way people integrate sensory information by using eye-tracking equipment and various physiological measures at the same time.

Another level of AugCog activity involves developing systems that can take account of human values through emotional feelings. Engineers at the MIT Media Lab are working on sensors and algorithms that recognize emotions like interest, alertness, stress, and fatigue. Direct measurement might be difficult; yet, unless these responses can be identified, soldiers in the field might be burdened with too much information by some of the new equipment being developed, like a visor on a helmet that displays messages from a communications center. It might be best to modulate the amount or type of information that is being sent, depending on the soldier's emotional condition. So, simple tests are being tried that are surrogates for more direct identification of mental states deep in the brain itself. For example, a pressure-sensitive computer mouse has been designed as a measure of frustration in performing a task.

The MIT team argues that the introduction of new technologies should be accompanied by some way of assessing how willing the soldier is to adopt it, an aspect the team considers a practical application of the ethical value of autonomy. This admirable recommendation needs some unpacking. Respect for autonomy is indeed a cornerstone of modern medical ethics, a principle derived from many philosophical and religious traditions that value the human person as a moral agent with intrinsic worth. But what role does respect for autonomy play in a military context? Soldiers aren't normally asked for their informed consent before accepting what their command regards as the best preparation for battle. If the question is a purely moral one, then it's difficult to see why these AugCog enhancements should be regarded differently from anything else a soldier can be legally ordered to do. But perhaps the MIT team is indirectly identifying a different issue: that this level of management of the perception and cognition of one human being by another is a sufficient departure from our usual respect for people that it needs to be appreciated in some way, partly by putting the warfighter in the decision-making loop. I'll come back to the ethics of brain-machine relations later.

DE-STRESS FOR SUCCESS

Stress under fire is a recognized cause of human failure under fire, and perhaps the principal one. But crisis also concentrates the mind, helping us to focus and adapt to new situations. Evolution has given us this way to augment our normal cognitive powers in a pinch. Reducing the distracting effects of stress while leaving the adaptive ones alone should improve cognition. Drugs have long been of interest in managing combat stress, as they have been in civilian life, but their effects are hard to control and can hamper the important focusing function that should accompany a scary situation. So rather than trying to change the capacity of the individual to deal with stress, it may be wiser to reconceptualize the problem as one that involves a system, one composed of a soldier and the equipment the soldier uses.

In this way, the burden of stress management is placed on this system. Either member of the system can take up the slack for which it is best equipped. Human beings have the capacity for creativity that computers generally lack (so far), while computers are immune to stress and can greatly enhance human ability to handle information and improve human attention. The mission of the DARPA research program "Improving War-fighter Information Intake Under Stress" is

> to extend, by an order of magnitude or more, the information management capacity of the human-computer warfighting integral by developing and demonstrating quantifiable enhancements to human performance in diverse, stressful, operational environments. Specifically, this program will empower one human's ability to successfully accomplish the functions currently carried out by three or more individuals.

As the mission description continues, "The main goal of this program is to develop a closed loop computational system in which the computer adapts to the state of the warfighter to significantly improve performance." Considering the rapid growth in processor speed and memory and the power, miniaturization, and "ruggedization" of hardware, it should be possible to find new ways to make the soldier and the warfighting tech-nology work together, which should "fundamentally re-engineer military decision making," according to the program Web site.

To get feedback from the soldier in the field, an ideal arrangement would be some sort of wearable device that would record and interpret brain activity. That information could then be transmitted in real time to a command center that could make appropriate adjustments to the fight-er's condition. Pretty interesting idea, and, as usual, DARPA is way ahead of us. Already there is a prototype helmet with sensors that record blood flow in the cortex down to five centimeters deep. In principle, any cerebral blood flow data that can be associated with a mental state—anxiety, stress, confusion, etc.—could be remotely monitored. Orders could be changed or information clarified without soldiers even needing to report accurate-ly or candidly on their states of mind, something that combat personnel might be reluctant to do on their own for fear of compromising their ser-vice record.

The projects involving cognitive-feedback helmets straddle Aug-Cog and "mind reading" technology, so I'm saving the details for anoth-er chapter. It's enough to note that these projects are consistent with the AugCog vision of creating continuity between brain and machine, in the process obscuring the boundary between artificial and "natural" intelligence. When we can't tell the difference, intelligence will have been truly augmented and the tactical possibilities will be immense.

Finally, it's worth emphasizing that these projects are looking toward the very long run. AI experts haven't reached the point where machines can understand their own programs, let alone interact meaningfully with the human mind, though given the rate of increase in computer power, some believe the moment of machine self-awareness isn't all that far away. It's a nice question whether, when that moment comes, the brain-machine relationship will in turn become exponentially more intimate.

"WE ARE THE BORG"

Earlier I mentioned the symbiotic relationship between people and computers envisioned by the AugCog project. The idea of human-machine "symbiots" will strike a chord with fans of *Star Trek: The Next Generation.* The Borg is a species of beings that were once organic but have assimilated mechanical and electronic parts. There are no true individual "borgs;" *they* are "The Borg," composed of units instead of separate selves, a collective that is something like an ant colony. The scary part is that The Borg's raison d'être is to assimilate into itself every organic being it can find.

I wish the analogy between AugCog and The Borg were original with me, but I have to credit it to several members of the Institute for Human and Machine Cognition. In a paper called "The Borg Hypothesis," Robert Hoffman and colleagues note that long-duration space travel poses severe and perhaps intractable challenges to organic creatures, such as the bone loss associated with a zero-g environment, or the severe radiation exposure. For a species intent on leaping off its home planet, the implications are clear. As Hoffman and his colleagues observe:

> What we are reaching for here is a new meaning of evolution. Geobiological evolution on Earth has yielded creatures (humans) that can reengineer their own physiognomy (for example, artificial limbs), their own anatomy (for example, cochlear implants), and even their molecular biology (for example, gene therapy). Through human-machine symbiosis, we are on an evolutionary threshold where our species is capable not only of deliberately affecting its own evolution but also of changing the rules by which evolution occurs.

The observation that human beings are changing the evolutionary rules is not a new one, as the authors acknowledge. Programs intended to reduce the incidence of genetic disorders such as Tay-Sachs and sickle cell disease change the course of evolution as well. Genetic counseling is high tech, but low-tech public health programs that seek to avoid or minimize epidemics obviously have a huge evolutionary effect as individuals who would not have otherwise survived do so and reproduce. Bioethicists have long debated the wisdom of interventions that not only avoid passing on heritable diseases but appear to select for certain characteristics, such as sex or, in the not-too-distant future, height or eye color. Some of this smacks of eugenics, but driven by parental choices rather than a tyrannical state founded on racial theory. In fact, the human brain seems designed to enter into its own evolutionary trends. Human intelligence does create itself in a way that sets it apart from even the relatively high intellectual endowments of other primates.

But these evolutionary effects are organic. The human-machine symbiosis envisioned creates a different set of stakes that reaches to what it is to be a human being. Are we on a slippery slope toward a transformation of the human being into a mere disposable unit of the Borg ant colony?

THE DIGNITY THING

The vast field of brain-machine interaction could easily be the subject of its own book, but let's try to get our philosophical bearings before moving on. What roborats, primates controlling machine movements with mental activity, and the integration of human intelligence with electronic devices have in common is that they are surprisingly, perhaps even shockingly, new ways to modify the brain's action. Although these examples raise lots of philosophical issues, one that stands behind all of them is whether they entail some sort of assault on dignity, even that of the lowly and largely despised rat. Many would find the idea of mine-clearing roborats to be infinitely preferable to endangering humans, especially considering all the children who have been maimed in minefields throughout the world. It's not at all clear how to ascribe "dignity" to animals, but those who do might not be persuaded. Adapting the technology to humans, even if it could be done (an idea that is well beyond the limits of the roborat experiments), wouldn't seem to provide any advantage over using gung-ho types. But for the sake of argument, let's say it did provide greater efficiency of operation and fewer casualties than the old-fashioned self-directed mine-clearing personnel. There is one consideration that might make remote-controlled human mine clearers ethically acceptable, if not palatable: suppose they gave full voluntary consent to the procedure, based on evidence that with someone else pushing their buttons, they are more likely to survive the dangerous task.

Even if this kind of practical advantage could be demonstrated and even if the mine clearers gave consent, many would find the whole idea of turning people into cyborgs, albeit temporarily and for good purposes, unacceptable. The crux of the problem is the idea that such programs would undermine human dignity and lead down a slippery slope toward humans turned into robots for bad purposes. It's very hard to assess how likely that kind of long-term outcome is, which is probably why the question, Is the whole brain-machine research program morally tainted? will excite very different responses among different people. But then, much the same could be said about all military research programs.

And then there is, again, the dual-purpose aspect of the research. If turning people into cyborgs to fight wars is wrong because it's an affront to dignity, then why exactly is turning them into cyborgs so they can manipulate artificial limbs a triumph of the human spirit and not an affront to dignity at all? To take more familiar examples, we already have cyborgs among us: heart disease sufferers with pacemakers and amputees with artificial limbs managed by electronic sensors. If the worry is that artificially augmenting our cognitive capacity is somehow different from repairing or enhancing the rest of our bodies, then we'd better start worrying about the Internet and the computer on which I'm writing this sentence, both ways to extend the reach of our cognitive powers. What if the electronic device that rides on my hip today would be more efficient if implanted in my brain tomorrow? Should brain-injured people with implanted microchips be regarded by the defenders of human dignity as all that different from pacemaker-dependent people? Or is correcting deficits caused by disease and injury one thing, but deliberate enhancement of "normal" functioning another? Is the latter an assault on human dignity, but the former mere healing? What if the "healing" turns out to make the patient stronger and smarter than she was before?

The problem might lie in the idea of dignity itself. We appeal to the idea all the time when we talk about ethics, but can it bear the weight? How does it clarify what we should and should not find acceptably human? The current President's Council on Bioethics quite properly worries about human dignity in its published reports, yet there is surprisingly little analysis of the concept in the philosophical literature. Years ago, I heard a Jesuit professor of theology at a prestigious university express doubts that the idea of dignity is clear enough to carry the philosophical burdens that modern ethics places on it, that it's too ambiguous. If we won't buy the behaviorist B. F. Skinner's suggestion that we should simply go "beyond freedom and dignity" (his notorious book title of forty years ago), what does dignity require or rule out?

To get some help, I spoke with Leslie Meltzer, a Yale-trained legal scholar who is also a religious studies Ph.D. student at the University of Virginia. Meltzer won just about every recognition the university can give while an undergrad and has a bright future as a scholar. When I spoke with her, she was writing her doctoral thesis on the idea of dignity, which, of all the human values there are, might be the most cited and the least understood.

The ambiguity of the meaning of dignity shoots through our debates about human values. During the 2004 presidential race, for example, candidates and commentators debated whether human dignity would be violated if human embryos were used in research, or whether permitting people to suffer without trying to use embryonic stem cells to treat their diseases would be more of an affront to human dignity. Since nobody had bothered to define dignity, neither side really knew what it was talking about. We've seen the same thing happen in the euthanasia debate: does helping people die, or at least easing their way, uphold human dignity, or is it a threat?

When I presented the brain-machine experiments to Meltzer, she observed that dignity's very "squishiness" as a concept is an advantage in controversies like those about death and dying. "It allows people on opposite sides of the issue to appropriate the word. It provides sort of a trump that people can use regardless of which side they're on, and in that way the ambiguity of the word is useful." Dignity, whatever it means, is too important to be left to partisans on either side of these difficult issues.

That said, how does it help us sort out our feelings about the roborat? Meltzer urges that we start by understanding the complicated history of the word.

> "Dignity" comes from the Latin *dignitas,* which meant to have worth, to have rank. In the medieval world it was used mostly for people who had nobility, which could be acquired by, say, service to the king. Though it could be passed on from father to son, at the time it was not something you were simply born with, the way the word is used now, but something you got because of your social relationships, a relational term originally. Before that, pre-Christian pagan thinkers like Cicero used it to connote virtue or honor. When the Latin was translated into modern languages like German, "dignity" became *Würde,* and, in French, *valeur,* which means value or worth. Only in the Enlightenment do we see the word "dignity" with the word "human" in front of it. That changes the word altogether, because then dignity is about equality and liberty, something that all people have, the inherent dignity of man.

Somewhere toward the later medieval period and before the Enlightenment, the English word "dignity" appeared. It referred to inanimate objects, including natural objects. Later on, Francis Bacon cited the "dignity of science," and there is literature from that time on the dignity of fruit trees, even planetary functions in the period of Galileo. Even as human dignity emerged as an important political concept, there was this undercurrent, at least in English, that seemed to qualify it for other realms. "So there's really an important dichotomy in the history of the word," Meltzer said.

And maybe this dichotomy is instructive as well. It helps explain why we can ascribe dignity to people with noble virtues who signify the best humanity has to offer but also to vermin like rats, not for their own worth but in terms of their place in the natural order. It's not about the rats themselves but something in our moral history embodied in our language that warns us to take care about mucking around with this framework, turning living things into quasi-machines, whether rats or people.

Yet the human brain does have that highly evolved capacity to remake itself, and the Enlightenment teaches a confident, positive attitude toward the future. The eighteenth-century philosophes were less focused on the intangibles of dignity (with all its ancient implications of noble birth) than they were on the concreteness of progress. I am one of those who believes that, by and large, the Enlightenment attitude has served us well, and that when humankind has gone astray in subsequent centuries, it was because we failed to keep in mind the value of openness and a critical attitude toward assertions about reality. Out of that tradition, and in a remarkably short time, has flowed the very neuroscience that we're talking about.

3:

MIND GAMES

❖❖❖

ALTHOUGH THIS BOOK IS MAINLY CONCERNED with the national security implications of "high-tech" brain science, there is continuity between the psychological theories of the past and the neuroscience of the future. During the cold war, American behavioral scientists received unprecedented levels of funding from private foundations and from the federal government. Much of this support went to research on individual and social psychology problems that emerged in World War II and Korea. Some of the most prestigious institutions in the United States and Canada were involved, though if the research was classified even the professors themselves may not have always known exactly the real source of their funding. In a country still emerging from a world war in which scientists had played an important role, financial support from national security and intelligence agencies didn't necessarily raise questions. For many of these important scientists, the government's financial support was for work that started during World War II and continued after it. I will have more to say about this legacy later on, but one aspect of the research done in the 1950s came to mind as the Abu Ghraib prisoner abuse scandal broke in May 2004, just days after I decided to write this book.

HUMILIATION NATION

It is now well known that, in the months following the U.S. invasion of Iraq, dozens of Iraqi prisoners were mistreated while in U.S. custody at

Abu Ghraib, the infamous prison complex formerly associated with Saddam Hussein's tyranny. Seymour Hersh's description of the incriminating photographs appeared in the *New Yorker*:

> The photographs tell it all. In one, Private [Lynndie] England, a cigarette dangling from her mouth, is giving a jaunty thumbs-up sign and pointing at the genitals of a young Iraqi, who is naked except for a sandbag over his head, as he masturbates. Three other hooded and naked Iraqi prisoners are shown, hands reflexively crossed over their genitals. A fifth prisoner has his hands at his sides. In another, England stands arm in arm with Specialist [Charles A.] Graner; both are grinning and giving the thumbs-up behind a cluster of perhaps seven naked Iraqis, knees bent, piled clumsily on top of each other in a pyramid. There is another photograph of a cluster of naked prisoners, again piled in a pyramid. Near them stands Graner, smiling, his arms crossed; a woman soldier stands in front of him, bending over, and she, too, is smiling. Then, there is another cluster of hooded bodies, with a female soldier standing in front, taking photographs. Yet another photograph shows a kneeling, naked, unhooded male prisoner, head momentarily turned away from the camera, posed to make it appear that he is performing oral sex on another male prisoner, who is naked and hooded.

Newspaper accounts indicate that humiliation was routine at Abu Ghraib as well as at the Guantanamo prison camp in Cuba, where detainees from the Afghanistan war were held. Soldiers and civilian consultants who were not directly involved in interrogations but saw naked prisoners assumed that it was part of the intelligence-gathering process, as was later confirmed by military officials in response to the Red Cross. Some of the civilian contractors filed complaints about the practice.

In the days and weeks following these revelations, various experts on the military and intelligence operations expressed their doubts that these acts were simply spontaneous expressions of sadism. Although the young soldiers were obviously poorly supervised, the International Committee of the Red Cross concluded that there was a systematic element to the abuse, part of an effort to obtain information about terrorist plots and the anticoalition "insurgency" in Iraq that arose after the fall of Saddam. The Army inspector general found no such pattern and concluded that the problem lay with a few individuals, a determination that was met with skepticism by both Democratic and Republican members of Congress.

The Third Geneva Convention of 1949 appears to rule out such treatment: "Prisoners of war must at all times be humanely treated. Any unlawful act or omission by the Detaining Power causing death or seriously endangering the health of a prisoner of war in its custody is prohibited, and will be regarded as a serious breach of the present Convention. . . . Prisoners of war must at all times be protected, particularly against acts of violence or intimidation and against insults and public curiosity."

Two Georgetown University legal scholars, M. Gregg Bloche and Jonathan H. Marks, in 2005 published a paper in the *New England Journal of Medicine* in which they described certain interrogation policies at the American military prison in Guantanamo Bay, Cuba. Documents from commanders indicated that the personal medical records of detainees were to be examined by experts for evidence of any psychological weaknesses that could be used in interrogations, and medical personnel were specifically instructed to convey any such information that came to their attention through their interactions with prisoners. Advice received from psychologists and psychiatrists in a behavioral science consultation team was used to pinpoint an individual's vulnerability to certain stresses, including humiliating sexual provocations and contemptuous mishandling of Islamic symbols like the Qur'an. Critics observed that the exploitation of private medical records and breach of confidentiality are clear violations of international medical legal standards and could expose our own people to similar medically or psychologically abetted mistreatment if they are taken as prisoners of war or hostages.

Even before the questions about psychiatrists' and psychologists' roles in this episode began to surface, I wondered if the humiliation more or less deliberately applied at Abu Ghraib and perhaps other sites since 9/11 such as Guantanamo was a more direct product of well-established U.S. intelligence doctrine than has been acknowledged, an approach that had its origins in the Korean conflict several years after the Geneva Convention was drafted. At that time, the air was thick with rumors that American POWs were being "brainwashed," having their minds erased and reconstructed so that they would both disclose sensitive information and perhaps be turned into unwitting Communist agents. All kinds of methods were theorized to be part of this process, including drugs, hypnosis, intimidation, and humiliation. The best-selling 1959 novel *The Manchu-*

rian Candidate made brainwashing part of popular culture. Though the subsequent movie was a box-office failure (its release the week of the 1962 Cuban missile crisis didn't help), the 2004 remake starring Denzel Washington suggests that American fascination with mind control and political psychology has taken on a new life as terrorism has become a central concern.

As a former Army interrogator told a television news network after the Abu Ghraib scandal broke, "Humiliation has been used in the interrogation process for quite some time." Wondering if "quite some time" referred to the Korean War era, I contacted a psychologist and a psychiatrist I knew who had expertise in this area. One had worked directly with intelligence agencies; the other had students who were doing so. They couldn't induce their colleagues who worked on interrogation techniques to talk to me on the record, but through these intermediaries I did receive confirmation that, in these behavioral scientists' view, my theory was correct: theories about and interest in humiliation as an interrogation instrument in the U.S. intelligence establishment dated back at least to the early cold war.

Experienced interrogators generally agree that physical violence results mainly in the prisoner's willingness to say anything to avoid more pain. Little useful information can be gained this way, or at least little that is useful without corroboration from other sources. Far more subtle psychological approaches are normally required. The CIA's training manual includes in its title the code word for the agency in the Vietnam era, KUBARK, and dates back to 1963. It exhibits a great deal of psychological sophistication and experience, passed down for at least a generation:

> The effectiveness of most of the non-coercive techniques depends upon their unsettling effect. The interrogation situation is in itself disturbing to most people encountering it for the first time. The aim is to enhance this effect, to disrupt radically the familiar emotional and psychological associations. . . . When this aim is achieved, resistance is seriously impaired. There is an interval . . . of suspended animation, a kind of psychological shock or paralysis. It is caused by a traumatic or sub-traumatic experience which explodes, as it were, the world that is familiar to the subject as well as his image of himself within that world. . . . At this moment the source is . . . far likelier to comply.

What the manual calls the feeling of being "plunged into the strange" was not the product of guesswork, but the result of intense psychological study by some of the country's leading behavioral scientists. The work began more than a decade before the KUBARK manual was published.

PSYCHOLOGY'S COLD WAR DEBT

To a great extent, modern psychology and social science were founded on the financial support they received from national intelligence agencies during and after World War II. It has been estimated that a third of American research psychologists were part of the war effort, including some of the most important names in the field. These close ties remained after hostilities against the Axis powers ended. In the early 1950s, nearly all federal funding for social science came from the military, and the Office of Naval Research was the leading sponsor of psychological research from any source in the immediate postwar years. The CIA found ways to support large numbers of Ivy League academics, often without the professors' knowledge, as its funds were passed through dummy foundations that often gave grants to other foundations. Measurements of personality and interpersonal relations were primary fields of study.

Much of the psychological work conducted during World War II was concerned with the conduct of individual fighters. One landmark was a study by Harvard professor Samuel Stouffer and associates published in 1949, *The American Soldier,* which examined the attitudes of ordinary soldiers toward their military experience. Previously, the attitudes of the rank and file had not been considered important. Another was S. L. A. Marshall's 1947 report, *Men Against Fire,* which claimed only about 15 to 30 percent of men actually fired their weapons in combat. Marshall's results were hugely controversial and influenced combat training for Korea and subsequent conflicts, with further studies devoted to identifying personality characteristics likely to enable an individual to function effectively and aggressively under stress.

Another area of growing interest to the post–World War II U.S. military, especially after President Truman desegregated the Army, was the psychological differences between races. When hundreds of Korean War soldiers were assessed for their fighting quality by consulting psychologists from academia, "Negroes" received far lower scores than did whites, and southern blacks performed worse than northern blacks. Another conclusion was that education was a more important factor for skilled white fighters than for blacks. Later studies in the 1950s of Marines and signal corpsmen showed similar results. The methods used in these top secret studies would not withstand scrutiny today. The racial data are dubious at best, since the raters' race and potential prejudices were not noted. Yet the net results of the studies of the psychology of warfighting, including data on correlations with such characteristics as birth order, physique, and even sense of humor, did significantly reduce the proportion of combat soldiers who were rated nonfighters in later years.

Academic psychology was ready when, in 1953, the legendary CIA Director Allen Dulles decided to make the management of foreign agents and soldiers more scientific. Receiving briefings about the abuse of allied POWs in Korea, he gave two Cornell University Medical School neurology professors the job of studying Communist brainwashing techniques. The resulting Wolff-Hinkle report was secret for many years. Contrary to the expectations of many who associated Asian interrogation with some exotic Oriental secret, the professors concluded that the success of Soviet and Chinese techniques resulted from a combination of exploiting human weakness and relentless psychological pressure.

The first phase was solitary confinement and subjection to demeaning and humiliating treatment by prison guards, who convinced the prisoner that there was no hope of release or contact with the outside world. There was also physical torture in the form of standing for long periods and being awakened repeatedly. After a few weeks, the prisoner usually broke down, and then the interrogation began. In a scene that could have been drawn from Dostoevsky's *Grand Inquisitor,* all of the prisoner's "crimes" from an entire lifetime were reviewed and dissected, including what were supposed to be his barbaric attacks on the innocent people of North Korea. When it appeared the ordeal might be over, it only started again. Finally, it became clear that only a full and signed "confession" would bring an end to the misery. The Soviet security service, the KGB, also used this protocol and claimed to have virtually 100 percent success with the process.

The Communist Chinese took the business one step further than the Soviets. After the "confession," the prisoners were placed in a group cell where they were lectured on Marxist and Maoist philosophy all day and criticized themselves for their ideological shortcomings. Any individual misconduct was a problem for the entire group, and all had to show that they were ardent believers in the party. It was George Orwell's *1984* come to life. Sometimes the results were life transforming, as though a religious awakening had taken place, a kind of born-again experience. Though apparently harder to achieve, this result reflected a more subtle understanding of psychology and group relations than that of the Soviets. To be sure, these softer measures were often combined with stress-elevating, spirit-breaking techniques such as lengthy periods in one physical position; control over urination and defecation; isolation; and sleep deprivation.

Dulles' CIA agents thought they might be able to do even better, but as Dulles himself said in a public statement, "We have no human guinea pigs to try these extraordinary techniques." That wasn't an insurmountable obstacle. A CIA front organization called the Society for the Investigation of Human Ecology funded the experimental work of the distinguished McGill psychiatrist Ewen Cameron, an American Psychiatric Association president and the first president of the World Psychiatric Association. Cameron may not have known that the CIA was indirectly supporting his work, but the agency was quite interested in his claim that he could in essence erase a mind and then reprogram it with new behavior patterns. Cameron's stated long-term goal was a treatment for schizophrenia he called "psychic driving," which involved bombardment with repeated verbal messages. The idea was to take certain emotionally laden "cue statements" from extensive interviews with the patient about his or her troubled life. In his 1979 book, *The Search for the "Manchurian Candidate,"* John Marks recounts the following message that was played on a tape loop sixteen hours a day for weeks:

> Madeleine, you let your mother and father treat you as a child all through your single life. You let your mother check you up sexually after every date you had with a boy. You hadn't enough determination to tell her to stop it. You never stood up for yourself against your mother or father but would run away from trouble. . . . They used to call you "crying Madeleine." Now that you have two children, you don't seem to be able to manage them and keep a good relationship with your husband. You are drifting apart. You don't go out together. You have not been able to keep him interested sexually.

When Cameron determined that this stage was sufficient, he switched to a positive message:

> You mean to get well. To do this you must let your feelings come out. It is all right to express your anger. . . . You want to stop your mother bossing you around. Begin to assert yourself first in little things and soon you will be able to meet her on an equal basis. You will then be free to be a wife and mother just like other women.

Cameron's experiments have been severely criticized as lacking plausibility as well as violating medical ethics, as I reported in my book *Undue Risk*. Here, I am less interested in assessing his theories and professional conduct than I am in noting the intelligence community's interest in them, including the references to sexual inadequacy in the negative phase of the experiment.

While Cameron was doing his work in Montreal, one of the coauthors of the report for Allen Dulles, Harold Wolff, a former president of the American Neurological Association, offered to do his own experiments in New York. Wolff thought that he could identify more effective interrogation and indoctrination methods than those the CIA utilized and asked for the agency's files on humiliation, among other approaches. He then took one hundred Chinese refugees and attempted to mold them into American agents who could be inserted back into the mainland. They were trained to withstand Chinese brainwashing by being "preconditioned." Partly drawing on that work, the American military developed a program called SERE—survival, evasion, resistance, escape—to prepare soldiers for this kind of treatment if they became POWs.

In 1956, Wolff reported the results of his indoctrination studies to the U.S. Group for the Advancement of Psychiatry. He observed that pain is most effective as a brainwashing technique when it is tied to hopelessness and humiliation. Wolff listed eight methods of coercion utilized by the Chinese Communists, including degradation. The failures of American soldiers in Korea to withstand interrogations compared unfavorably to the performance of Turks who were also held but who maintained their self-discipline, cared systematically for ill comrades, and always had a leadership hierarchy. Besides improved self-discipline, American soldiers were said to need more education in democracy and multiculturalism.

This point continues to have resonance in the post–cold war world, as the offending soldiers at the Abu Ghraib prison have been said to have inadequately understood not only Islamic Arab values and beliefs but also the democratic principles they were there to uphold.

Wolff's goals were still grander than Cameron's, aiming not only at advances in abnormal psychology but, as he told the CIA in the early 1950s, to understand "how a man can be made to think, 'feel,' and behave according to the wishes of other men, and conversely, how a man can avoid being influenced in this manner." Wolff's statement might sound pompous to the modern ear, but was in fact far ahead of its time, for he aimed to bring together various disciplines to study man's relationship to his environment, which he called "human ecology." That kind of language and synoptic theoretical ambition only became more widely familiar fifteen years later. Today, fields such as psychobiology and, of course, neuroscience similarly seek to bring disparate disciplines together to focus on topics that are inherently multidisciplinary.

HIGH ANXIETY

More than fifty years ago, the intelligence community was committed to advancing the cutting edge of science. Academicians were intensely interested in personality structure in those days. A Harvard study aimed at psychic deconstruction by humiliating undergraduates and thereby causing them to experience severe stress. As described in Alston Chase's *Harvard and the Unabomber,* one of the subjects was a young undergraduate named Ted Kaczynski. After Harvard, Kaczynski earned a Ph.D. in mathematics from the University of Michigan, then taught briefly at the University of California, Berkeley, after which he dropped out of society. For eighteen years, using homemade explosive devices, he terrorized those he viewed as agents of antihuman technology, especially anyone associated with universities or airlines. By the time he was arrested at his remote Montana cabin in 1996, Kaczynski left behind a trail of mayhem that included at least two murders.

Many agree that Kaczynski is both very smart and very crazy, though they might disagree about the exact proportions. Before his arrest, he demanded that the *Washington Post* and the *New York Times* publish a

35,000-word manifesto called "Industrial Society and Its Future," a document that expressed his philosophy of science and culture. His major target was the Industrial Revolution, the font of human enslavement, according to Kaczynski. "The system does not and cannot exist to satisfy human needs," he wrote. "Instead, it is human behavior that has to be modified to fit the needs of the system." The only way out is to destroy the fruits of industrialization, to promote the return of "WILD nature," in spite of the potentially negative consequences of doing so, he wrote.

Chase argues that Kaczynski's antitechnology fixation and his critique itself had some roots in the Harvard curriculum, which emphasized the supposed objectivity of science compared with the subjectivity of ethics. During Kaczynski's sophomore year at Harvard, in 1959, he was recruited for a psychological experiment that, unbeknownst to him, would last three years. The experiment involved psychological torment and humiliation that could have left deep scars in some of its subjects, according to some of the experts Alston interviewed.

When I first read about the Harvard experiment in an except from Chase's book in the *New Yorker* magazine, it wasn't only the experiment that got my attention. It was also the man who did the experiment, the brilliant and complex Harvard psychologist Henry A. Murray. Murray was one of my father's closest friends. Though his fame has diminished since his death, Murray was among the most important scientists of his day, the pioneer of personality tests that are now a routine part of industrial management and psychological assessments. It is not too much to say that contemporary psychology would be far different without his contributions.

In the late 1940s and early 1950s, Murray brought groups of students to my father's psychiatric hospital and training center in the Hudson Valley for weekend explorations in group psychotherapy. A letter from Murray to my parents upon the occasion of my birth hangs on the wall in my study at home. Reading Chase's account of Murray's experiments on America's homegrown terrorist philosopher, I sat in shock for several minutes, then I grabbed the telephone and called my octogenarian mother, who knew Murray well, or thought she did. "You'll never believe this about Henry Murray," I said. "Oh, my God," she replied. "We never knew anything about that." Later, Chase and I talked and corresponded about his discovery, and my mother was able to confirm some elements of Murray's personal life, such as a long-standing affair and his preoccupation with the mysticism of *Moby Dick*. But my mother and I were both astonished at the nexus of the director of Harvard's psychological laboratory and America's notorious Unabomber.

Henry Murray was a native New York blue blood who became a Boston Brahmin. He attended the finest schools, Groton and Harvard, and earned an M.D. from Columbia and a doctorate in biochemistry from Cambridge University. He dropped medicine and natural science for psychology after reading Carl Jung, publishing a landmark work in 1938 called *Explorations in Personality*. Before World War II, the U.S. government asked him to do a psychological profile of Hitler, and during the war he helped the Office of Strategic Services (later to become the CIA) to assess its agents. In the 1950s, Murray's personality test, the thematic apperception test, or TAT, was used to screen Harvard students who were then given LSD. This was one of the many CIA-sponsored experiments under contract with major research universities. Murray himself is said to have supervised psychoactive drug experiments, including those performed by Timothy Leary. According to Alston Chase, Leary called Murray "the wizard of personality assessment who, as OSS chief psychologist, had monitored military experiments on brainwashing and sodium amytal interrogation." Chase reveals that Murray was addicted to amphetamines and tried various hallucinogens.

All this was itself a lot to digest. I remembered Murray as the paradigm of the dignified Ivy League professor, tall, handsome, urbane, and reserved. The last time I saw him, in 1968 over breakfast with my parents in Boston, he wore a three-piece tweed suit. Still, the new information about his secret bohemian lifestyle, though a surprise in its details, my mother and I could accept. The experiment he performed on Kaczynski and twenty-one other students was another matter. It did not meet the ethical standards of the day, which, although less developed than they are now, emphasized carefully weighing and minimizing the risks of an experiment. Risk assessment is especially important when the subjects have little to gain from being part of the research. By any reasonable measure, the Murray experiment flunked that test.

The decade of the 1950s is widely stereotyped as oppressively if comfortably boring and conformist. While that might be true, there was a great deal of interest in alienation. Popular culture included films such as *Rebel Without a Cause,* and "beatniks" who challenged the public imagination. Social scientists also found nonconformists fascinating. They wanted to know what made these people tick. The personality tests that had been devised often showed that these subjects had rejected "normalcy," especially insofar as that implied an acceptance of the technological innovation that most Americans often uncritically embraced.

Subjects in the Harvard experiment were chosen from dozens of candidates who were screened for degrees of alienation from society—some quite negative, some at the other end of the spectrum, and some in between. Data on each of the students were coded to protect their privacy, with Kaczynski getting the name "Lawful," which Chase suggests might have been an ironic recognition of the potential for chaos the great psychologist could have perceived in this mild-mannered good boy.

Murray described the experimental process in a paper called "Studies of Stressful Interpersonal Disputations," published in the *American Psychologist* in 1963. The detached intellectual tone of the paper's title fails to capture the quality of the experience, which started blandly enough when the students were given a month to "write a brief exposition of your personal philosophy of life, an affirmation of the major guiding principles with which you live or hope to live." Then, "you and a talented young lawyer will be asked to debate the respective merits of your two philosophies."

When the day came, the subjects were taken to a room with bright lights and a one-way mirror. Electrodes were attached so that heart and pulse rates could be recorded, and the event was filmed. Despite Murray's published report, as Chase discovered from Murray's progress notes, the students were told in advance that they would debate another student, not a lawyer. The aggressive law student was sprung as a surprise, and he was instructed to attack the student. The surprised subject typically tried to defend himself, and he became angry at having his personal philosophy so harshly criticized. In fact, the whole scene was calculated to excite the emotional and physiological responses associated with a demeaning and even threatening situation—being strapped in under large white lights with a camera rolling in front of a glass in which only shadowy figures appeared. The students reported feeling warm, started sweating with hearts pounding, and shifted around in obvious discomfort.

Later, they were asked to watch the film of themselves being verbally deconstructed. For the young men who took pride in their intellect and self-presentation, seeing themselves so frustrated and inarticulate was quite disconcerting and undermined their self-image. If the event was mainly an example of stimulating stress, this part was truly humiliating for an Ivy League student in the 1950s. This was a different world from today's confessional culture in which people disclose their private shame on *Jerry Springer;* the loss of poise and self-control was deeply disturbing to these young men. Did the whole experience create the Unabomber? Unlikely, but it certainly didn't make Kaczynski more sanguine about science. For his part at least, Kaczynski told one of his lawyers that the Murray experiment was "a highly unpleasant experience."

MIND CONTROL

The brainwashing in Korea that so concerned U.S. authorities was sometimes combined with the use of new drugs, or new uses of old drugs, to alter consciousness. "Mind control" with the help of drugs like LSD and mescaline was thought by some to be a useful first step before brainwashing itself. The irony is that only a few years later, these substances became the centerpieces of a movement intended to release people from the constraints of the "establishment," to liberate minds brainwashed by grubby capitalism and war fever.

Marks' *The Search for the "Manchurian Candidate"* and a raft of other books published in the late 1970s about hallucinogenic research by the Army and CIA were based on government revelations that were themselves only possible in the post-Vietnam, post-Watergate era. For a generation that had grown up thinking that their drug trips were rooted in a quest for innocence and purity, Marks' history was disconcerting. At the beginning of his book, he noted that the accidental discovery of LSD by a Swiss drug company scientist working with derivatives of ergot took place in 1943. In the same year and only two hundred miles away, experiments with mescaline were being done by SS doctors at the Dachau concentration camp. The Nazis, though, pronounced themselves disappointed with the mind control potential of mescaline as an alternative to traditional interrogation techniques.

Marks summarized the information gathered by Idaho Senator Frank Church as part of a congressional investigation into inappropriate activities of intelligence agencies and the military during the 1950s and 1960s. These activities went well beyond sponsoring the Cameron experiments in Montreal and included dosing both CIA personnel and private citizens with LSD. Experiments done by the Army Chemical Corps resulted in the death of a hospitalized psychiatric patient in New York, and LSD exposures involved thousands of soldiers. The CIA's anthrax expert, Frank Olson, was given LSD and subsequently fell to his death from a New York City hotel room window under circumstances that remain suspicious. The backlash against the CIA after these revelations has been blamed for subsequent intelligence failures that led to 9/11. Members of Congress have called for a return to a more aggressive approach to intelligence that includes getting involved once again with unsavory tactics and individuals. So the pendulum swings.

Intelligence agencies were interested in hallucinogens to determine both their potential in aiding interrogations and their usefulness as nonlethal weapons to disorient enemy units. Although the drugs never turned out to be satisfactory for these purposes, some might argue that the experiments weren't allowed to run their course because of the public outcry. Similarly, I have spoken to senior psychiatrists who still think that some form of LSD could be useful in psychotherapy but that we may never know because of the stigma left over from the Timothy Leary days.

Neuroscience is, however, learning a great deal about naturally occurring substances in the brain that may open up new vistas for influencing subjects of interrogation. Chemicals in the brain associated with moods and emotions are being identified. What if they could be exploited? A group at the University of Zurich gave volunteers either a spray of a hormone called oxytocin or a placebo. They then ran the subjects through a decision-making exercise in which they had to choose whether to invest assets through a "banker." In 2005, the scientists reported that those who inhaled the oxytocin were more likely to trust the banker and take a risk. The apparent effect lasted about two hours.

If these results bore out, security agencies would surely find the concept intriguing as a way of enhancing the trust of suspects, especially when time is short. Imagine that a bomb threat has been received from a terrorist organization and an individual in custody is believed to have information. A dose of trust hormone could help persuade the subject that the interrogator is sincere in promising that sharing details now will make for better treatment later.

However, one of my neuroscientist consultants expressed skepticism about the report from Switzerland due to a simple physiological fact: oxytocin sniffed does not get past the blood-brain barrier. The hormone would also be degraded as it attempted to make its way from the nose to the brain, which would take hours. He attributed the Swiss experimenters' "results" to coincidence.

So it seems there's more work to be done to achieve a trust drug and, anyway, working out the delivery mechanism would pose another challenge. The effect could not easily be exploited by people who wanted to gain the trust of their subjects, such as politicians and marketers, whose activities probably stimulate the production of oxytocin in their audiences anyway.

There were other innovative approaches to influencing the mind during the cold war that did not rely on chemicals. Since the 1970s, there have been reports about Soviet and Chinese interest in "psychotronic" weapons intended to influence psychological and physiological processes at a distance. One of the proposed avenues to other minds has been electromagnetic radiation or "extremely low frequency" (ELF) waves. American interest in these matters was partly a response to Soviet activity. With the collapse of the Soviet Union, it is an open question whether national security and science agencies will continue to probe all the possibilities presented by neuroscientific advances, including interventions that might be considered attempts at mind control. As evidence that such efforts will be renewed, human rights advocates claim that references to mind control or psychotronic weapons, including summaries of information about Russian and Chinese efforts, remain classified.

Although psychotronic warfare has been seized upon by those who believe a security agency is controlling or disrupting their brain, its goal as information warfare would be to attack communication systems, thus causing a catastrophic infrastructure failure. Jamming transmissions by

Saddam's radar installations in the run-up to the Iraq war was an elementary example of such tactics. Similar principles might be applied to the mental energy of the warfighters themselves, perhaps by "pulse-wave weapons," which would disrupt motor signals from the central cortex. Once again, though, reports about Russian possession of such weapons are highly disputed, let alone the technical capabilities the weapons might have.

Perhaps more within reach are developments in functional magnetic resonance imaging (fMRI) technology. The advent of fMRI has been a boon to neuroscientists interested in correlating blood flow with just about every imaginable human experience. If the basic mechanism could be improved to detect blood flow some distance from the target brain—much as satellite images detect the temperature of objects on the earth's surface—it would be possible to install surveillance systems in sensitive public spaces such as airports. Individuals with increased blood flow in neural systems associated with aggressive behaviors could be singled out and stopped for questioning. As air travelers have noticed in the past few years, whether this approach would provide a security benefit could be beside the point for authorities eager to appear to be doing all they can to protect the public.

But like so many potentially scary technologies, this one presents more technical difficulties than might be apparent. A magnet powerful enough to generate images at a distance would also suck anything containing iron into its grasp. As New York University neuroscientist Paul Glimcher pointed out to me, with current fMRI devices applied to this kind of use, any metal of that type you were wearing or carrying would fly off your body or out of your clothing as soon as you came within range. If you had a pair of pliers in your pocket, you could end up pinned against the machine and possibly quite badly injured. The flying public would surely prefer even the inconvenience of the current system with its occasional pat-down to being pasted to a magnet on the way to the gate.

BACK IN THE USSR

Competition with the Soviets was surely one reason for the American intelligence community's interest in parapsychology. These kinds of studies, while always somewhat marginal in American science, achieved virtually mainstream respectability in the Soviet Union. Anything that might contribute to the international revolutionary struggle was highly valued. And if a field of study didn't contribute to that struggle, there was something wrong with it. Much of Soviet science was hopelessly politicized, especially evolutionary biology, which was obliged to follow Marxist principles. Instead of the concept of natural selection (giraffes got long necks because the animals with longer necks were able to reach fruit on tall trees and survived to breed), it was taught that change could happen in a single generation. Otherwise, the proletariat could not become qualitatively transformed into a revolutionary class as was required by Marxism. Many promising careers were destroyed by this patently false notion, and the life sciences were held back for generations.

Nothing, however, compared to the overt abuse of psychiatry in the service of the Soviet state. Many psychiatrists and psychiatric institutions became de facto political operatives, finding dissidents to be stricken with psychopathology that required long-term institutionalization. The intentional misdiagnosis of dissidents was of great concern to the psychiatric community in the West. For years, organizations such as Médecins Sans Frontières and the World Psychiatric Association worked on behalf of reform, but they were strenuously resisted by the Soviet All-Union Society of Neurologists and Psychiatrists. A 1989 American Psychiatric Association delegation found that out of a sample of twenty-seven patients, twenty had been hospitalized for questionable reasons.

In the Soviet Union, psychiatric diagnoses were a very useful national security tool, an experience that should serve as a cautionary note as neuroscience becomes ever sophisticated in identifying the neurologic basis of differences between individuals. We have already seen how the brain sciences can be useful in fighting wars and in conducting intelligence operations. An item on the totalitarian wish list, controlling domestic dissent, can also benefit from refinements in understanding and managing neural processes. Though the Soviet Communists took a typically ham-handed approach, future tyrannies will likely be less clumsy.

An authority on the Soviet political system, Theresa C. Smith, has identified certain peculiarities in the psychiatric culture of the former USSR. The definition of mental illness was much broader than in Western countries, enabling noncomformists to be labeled as "asymptomatic schizophrenics" and hospitalized; in the West, schizophrenia must be active (involving, say, hallucinations) to justify even the diagnosis, much less the hospitalization, of patients. Also, in an inheritance from old-fashioned eugenics, the Soviets emphasized genetic causes of mental illness, so that whole families could be labeled insane. Seeming to be healthy was no protection, as that might just be a case of "dissimulation" (lying), which was supposed to be typical of certain "paranoids."

Another important factor, according to Smith, appears at first not only harmless, but a positive good: the regime made sure that local medical clinics offered psychiatric services. Surely, access to this kind of care is a good thing. The trouble is that in a repressive political system, these local dispensaries could turn into outposts for officials to monitor suspicious individuals. Based on dubious psychiatric grounds, millions of nonconformists were put on a register that was used by police agencies for occasional roundups.

It appears that drugs were also used on people accused of political crimes. Used in very large doses, antipsychotic medications were a way to punish those who complained or broke the rules, or to force "confessions" of guilt to crimes, thus ruining a dissident's credibility. Punitive dosing is hard to prove because a physician can always say that a high dose was necessary in a particular case, and, as Smith reports, the overall drug dosages were high in Soviet psychiatry. But some drugs that have no known benefit were also used, including sulfazine, which can cause pain and muscle spasms. The more specific connections can be made between drugs and their effects on the brain and nervous system, the more vulnerable future political rebels could be to a dictatorship that determined to fuse medicine with its own survival.

THE PROPAGANDA WARS

Still more mundane than the application of psychiatry for political purposes is old-fashioned public relations. Compared with LSD, or even experiments with someone such as Ted Kaczynski, dropping leaflets over enemy territory is pretty boring stuff. But in a book about the brain and national security, I would be remiss if I didn't point out that propaganda of all kinds is basically an attempt to influence the minds of soldiers and citizens in the opposing camp. Advertising is a direct, simple, cheap, and nearly universal mind control technique. It also might be the psychological tactic in which the United States and other major powers have invested the most. Whether it works as a weapon of counterinsurgency is another question.

The "science" in low-tech psychological warfare concentrates on identifying the vulnerabilities in the target population's social psychology. Examples are ethnic rivalries that can be exploited or dissatisfactions with specific aspects of the regime. Hints about the most promising approaches can be gleaned from diverse sources, such as the way that government-controlled radio emphasizes certain national achievements rather than others (hyping corn production may be a sign that corn has been in short supply in recent years), or even frequent themes found in graffiti. Similarly, when individuals are to be indoctrinated, it's important to know their individual vulnerabilities, such as their conviction that they have been passed over in promotion for less qualified but more politically connected rivals. During open conflict, defectors have been less ideologically committed than other soldiers. Those considering defection must have the sense that they will be well treated and that they will be asked only to volunteer information, not forced to provide it. Once they make the decision to defect, they tend to be very cooperative.

Both the United States and the Vietcong engaged in aggressive propaganda campaigns during the Vietnam War. Vietcong leaflets denounced South Vietnamese government officials as lackeys of the imperialists and the American soldiers as bloodthirsty racists. The United States dropped hundreds of millions of leaflets urging insurgents to return to their homes rather than continue to endure the hardships of life as guerrilla fighters. Many leaflets were dropped a few hours after B-52 strikes warning that there was more bombing to come. Leaflets directed at the North Vietnamese soldiers noted the likelihood that if they were killed they would be buried far from home, a culturally significant concern.

The famous "deck of cards" of important leaders in Saddam's regime was only one of many written items intended to influence Iraqi thinking. U.S. Central Command leaflets dropped during the 1990s, when the "no-fly" zone was being enforced, discouraged soldiers in tanks or artillery batteries from firing on coalition planes, warning they would risk being destroyed by overwhelming force. As the ground assault got under way in 2003, notices were air-dropped announcing the defection of thousands of Iraqi soldiers and stating that the coalition's goal was the defeat of Saddam rather than to harm the Iraqi people. On my bookcase I have a matchbook distributed in Afghanistan as and after the Taliban regime was deposed. The front is adorned with a photo of Osama bin Laden and a pile of gold coins with the inscriptions "YOU DELIVER WE PAY" and "REWARD."

Small print over the flap reads: "Be safe—keep cover closed." Even bounty hunters shouldn't play with matches.

HAS HUMILIATION OUTLIVED ITS USEFULNESS?

From the CIA's worries about Chinese brainwashing to mind control to LSD to the abuse of Soviet psychiatry to the Abu Ghraib prison scandal, it's been a strange half-century trip. On the whole, attempts to use knowledge of the mind for security goals have had a few isolated successes (the use of personality inventories) but largely resulted in a lot of wasted money on what appear to have been outright clunkers (mind control, LSD, etc.). Other low-tech measures such as propaganda leaflets were cheap, carried low risk, and might sometimes have contributed to pacifying the locals. These approaches to the management of human thought and action not only were based on cold war science, they were pursued in the context of cold war politics. But both the science and the politics have changed, as the Abu Ghraib prison scandal illustrates.

Clearly, the behavior in the Iraqi prison went far beyond "legitimate" interrogation tactics and seems to have sunk into frank sadism, but the affair focused more attention on U.S. intelligence policies than had been the case since the 1970s. These policies are still rooted in the advice of the cold war psychologists, who developed protocols that include humiliation and stress. Although less discussion has taken place about the techniques used in Afghanistan and in other Iraqi prisons compared with the high-profile cases of Abu Ghraib and Guantanamo, my sources as well as published accounts indicate that the same general approach was applied at those other detention centers, though apparently without the extreme misconduct at Abu Ghraib.

Thus it was no accident that the first images of a captured Saddam

Hussein displayed him in the most humiliating circumstance short of pornography: having his orifices checked for infestation by a medical worker. Like the Abu Ghraib photos, the display of those images was likely a violation of the Geneva Convention. Yet we are told that Saddam has given little useful information, and the display of the once-feared Lion of Baghdad as a degraded street bum did nothing to quell the fury of dissident Iraqis.

As a professor of medical ethics, when I first saw the astonishing images of Saddam, I was immediately struck that the individual performing the examination seemed to be a competent health care professional. In that case, he or she might have been at risk of violating professional ethics. International protocols forbid publicly humiliating prisoners of war, and various codes state that doctors in particular are not supposed to participate in such activities and are in fact supposed to protect those who are ill or may be victimized. The same point applies to any health professionals who were in attendance at the Abu Ghraib prison: to the extent that they were aware of the abuse, they had a specific ethical obligation by virtue of their professional status to report it. The American Medical Association has expressed its concern about the possible failure to act of any doctors who were present.

Quite apart from the origins, efficacy, and ethics of human degradation as an instrument of warfare, the question of its political symbolism is also important: the implications of humiliating soldiers and clandestine operatives in an ideological conflict like the confrontation between capitalism and Communism are far different than they have become in what militants portray as a "clash of civilizations." Unlike radical Islamists, Russian and Chinese agents did not recall a thousand-year legacy of conflict with the Christian West. In particular, the Soviet Russians were Europeans and culturally Judeo-Christian, albeit religiously repressed. The Communist Chinese adopted a political philosophy pioneered by a European philosopher-journalist who was born a Jew, Karl Marx. And though capitalists and Communists disagreed in their interpretation of progress, at least they believed in it. Both worldviews were products of the Western Enlightenment, they were intellectual siblings, whereas militant Islam identifies with the world of the first millennium. Rather than provide a tactical advantage, in the conflict in which we are now engaged humiliation might serve mainly to feed the extremism at its core.

In this new and challenging policy environment, when the ultimate

Western fear is the takeover of a nuclear-armed state by militant Islamists, any advantages conferred by science will be as intensely sought by national security agencies as they were during the cold war. In the neurosciences, however, the basis for influencing other minds could be far more sophisticated than the likes of parapsychology, and far more precise than personality theory or propaganda leaflets. Blunt instruments like humiliation might go the way of the dodo bird. Perhaps neuroscientific breakthroughs will not only provide new opportunities to gather intelligence and develop new offensive and defensive capabilities, but help us to predict the behavior and understand the motivations of those arrayed against us in the twenty-first century.

And, perhaps, control them.

4:

HOW TO THINK ABOUT
THE BRAIN

THE EARLY 1950S THROUGH THE MID-1970S represented the preneuro-science era of psychological warfare. National security officials were main-ly dependent on the social psychological theories and pencil-and-paper psychometric tests developed by Henry Murray and others that emerged around World War II. The application of these efforts to military uses be-came known as "psyops" or "psychological operations." The first Depart-ment of Defense Worldwide Psyops Conference took place in 1963. In that year the Army also funded a project on "human factors and non-material special operations research."

PSYOPS

Peter Watson reported in his 1978 book, *War on the Mind,* that in the midst of this period, 1965, seven psychologists produced a classified paper for the Pentagon called *Phenomena Applicable to the Development of Psychological Weapons.* Only seventy pages long, the wide-ranging document covered ways that arms makers could enhance the psychological effects of new weapons, how fear could be enlarged, how perceptions could be al-tered, how the stress of combat could be increased, and how the psycho-logical differences among racial groups could be exploited. At the same time, another group of psychologists at the Human Resources Research Office (HumRRO) in Alexandria, Virginia (which had done "panic studies" of soldiers deployed to the atomic test shots in Nevada), suggested that combat units' shooting effectiveness would increase by half if the men at the end of the firing line had fewer bullets than those in the middle, because those at the end were more likely to be "trigger happy" and those in the middle would be more inclined to preserve ammunition. Another team of psychologists and anthropologists at Wright-Patterson Air Force Base found that different races found different odors offensive. Perhaps "smell bombs" could be devised to flush out guerrillas in the dense growth of Southeast Asian tropical jungles.

As the 1960s unfolded, an even more exotic notion of psyops (attempting to take advantage of what advocates of parapsychological phenomena call "psi") emerged as part of the "human potential" movement that also spawned sensitivity groups. A manifestation of the popular culture sweeping much of the developed world, the human potential movement included a vast array of beliefs and practices, including novel forms of psychotherapy, spirituality combined with psychedelic drugs, and a commitment to the study of parapsychological phenomena. Some of the scientists who were supported by defense agencies in the 1950s in less spectacular projects moved into these attempts to chart the expansion of consciousness, often with government funding. The leading academic parapsychology researcher, Duke University's J. B. Rhine, did ESP research with CIA support starting in 1952, a fact he revealed two decades later. Though Rhine was a pioneer, nearly everyone seems to have been carried along on the tide of the psychedelic sixties, even the cold warriors in the espionage agencies.

One of the principal observers of this period was John Wilhelm, a longtime friend and former neighbor of mine in Washington. John had a ringside seat on the sixties. A Princeton alum with a physical science background, John was an intelligence officer in the Navy through the Cuban Missile Crisis. He conducted briefings for field commanders, and was subsequently a science correspondent for *Time* magazine, where one of his early assignments was covering the manned space program. John also spent a year in Vietnam for *Time* in 1968–69. Returning home to catch the wave of social change, John expanded his coverage to examine some of the New Age's scientific culture (much of it pseudoscientific, he says).

When he started, he couldn't have known that his military intelligence background would also provide excellent preparation for his new beat, culminating in his period classic, *The Search for Superman* (1976). Wilhelm wrote about his encounters with famous alleged psychics of the era, including Uri Geller, who was said to bend spoons through the force of mind alone. By the end of the book, he was led into the literally spooky world of the CIA's interest in "remote viewing," the supposed ability to "see" places one has (presumably) never been.

I interviewed Wilhelm, now a prize-winning filmmaker, in his Washington, D.C., home in the late summer of 2004. Clearing the cobwebs of memory, he recalled first hearing about Geller from a former astronaut, Ed Mitchell, whose religious experience on the lunar surface moved him to start his own center, the Institute of Noetic Sciences, in Palo Alto, California. Wilhelm's attention was quickly drawn to the Stanford Research Institute (SRI) in Menlo Park, California, which conducted some of the experiments into paranormal phenomena, including the supposed powers of Geller and others.

Much of the work at SRI was secret, but enough information was publicly available that appeared to confirm a serious interest on the part of intelligence agencies to determine what was going on with these "paranormals," if anything. Wilhelm reported on an SRI experiment with a well-known psychic named Ingo Swann and a businessman named Pat Price, both of whom were allegedly able to "look at" a site just by knowing the geographic coordinates. An intelligence community monitor selected some coordinates and transmitted them in code to SRI. A CIA officer decoded the message and presented them to the SRI researcher, who then gave them to Price and Swann. Swann sketched an island in the South Indian Ocean with what was considered to be reasonable accuracy. Price gave a highly detailed description of an underground military installation somewhere in Virginia.

Price was then asked to "return" to the Virginia site and obtain code words. After seeing what Price reported, one security agent bemoaned, "Hell, there's no security left." But most intelligence officials and CIA psychologists would not accept the evidence. Wilhelm was told that the site in question was a station that obtained signals from Soviet satellites. An investigation into security leaks was initiated.

During the 1960s and 1970s, various government agencies paid for parapsychological studies, including DARPA, the National Institutes of Health, the Navy, and the CIA. At the same time, the Soviets invested in similar research, perhaps even more heavily, often under the heading of "psychotronics." Parapsychologists might not posit an explanatory theory, but the proponents of psychotronics contend that minds can interact based on psychic energy and also that electronic devices can influence psychic energy. Theirs is an attempt to subsume psychic phenomena under natural processes. The idea is that lower-frequency beams such as microwave radiation, which are at the other end of the energy spectrum from X-rays, can affect brain cells and thereby alter psychological states. The low-frequency bombardment of the U.S. embassy in Moscow by the KGB in the late 1970s seemed evidence that the Soviets were serious at least about exploring the possibilities of low-frequency weapons, trying perhaps to cause psychological problems among diplomatic personnel. A technical debate then ensued about whether it was possible for such energies to cross the blood-brain barrier, a protective wall formed by the vessels that carry blood to the brain.

Although this question has never been conclusively settled, psychotronics still has its advocates, a minority of whom contend that illicit experiments involving electromagnetic fields are being conducted by intelligence agencies. But the heyday of enthusiasm for such possibilities in the intelligence community seems to have passed over twenty years ago, when a retired Pentagon analyst and Army officer named Thomas E. Bearden attributed various events like Legionnaire's disease, UFOs, and mutilated cattle in the Midwest to Soviet psychotronic experiments, according to journalist Ronald McRae. But the apocalyptic weapons the Soviet Union was said to be prepared to release did not save the empire, and no such weapons of mass destruction were found during or after the cold war.

On the face of it all, this activity around psyops looks like evidence of serious interest on the part of both cold war superpowers. But Wilhelm isn't so sure. "This is a very murky area," he told me. "Even after years of looking at it, I can't be sure that all this wasn't for disinformation." In other words, although true believers get excited about this government activity—surely it means something if top security officials are committing money to studies—it could all have been to throw the other side off the trail and make them waste time and resources. It may be significant that the CIA closed the remote viewing program in 1995, with a report that concluded the results were disappointing. Would the program have been shut down if the Soviet Union were still in business? And what would an answer to that question mean?

Back in 1977 when Wilhelm went to the site indicated by the coordinates Pat Price supposedly remotely viewed, he found nothing but an open pasture. A Navy project manager overseeing a different SRI project suggested various possible explanations—that the coordinates the intelligence officer provided were wrong, that Price "saw" a nearby satellite communications center in West Virginia, or even that Price read the intelligence officer's mind.

BEWARE OF DUALISM

But mind control and psyops experiments are akin to trying to ride a bike that hasn't been assembled. That's where modern neuroscience comes in. The big issues in neuroscience have been surprisingly stable since the ancient philosophers and physicians first started working on them. One of these issues is where the mind is located, and this brings up an important point about our language: although "mind" can be used as a verb ("Mind your manners!"), "brain" cannot (except for the colorful but archaic threat, "I'll brain you!"). Ordinary language may be telling us something important about the way we think about the mind, that it is *functional* as well as—or perhaps rather than—a discrete object. We have to avoid completely identifying the mind with the brain. The mind also includes sensations delivered from elsewhere in the nervous system some distance from the brain itself, so we need to take care not to collapse the two.

The trouble lies in conceptualizing how the mind and the brain can coexist. Historically, there have been two major thoughts on this, but science is beginning to lend an opinion that might help reconcile them. The first line of thought is materialism (which in the twentieth century took the form of behaviorism and Skinnerism in the psychological literature). First posited by Lucretius during the time of the Romans and picked up by thinkers such as Spinoza, materialism's basic argument is that the mind results from the sum of the brain's parts. Every thought, movement, feeling, and emotion is a direct result of some physically identifiable action in the brain.

Interestingly, a passage ascribed to the Greek physician Hippocrates from around 400 BC caught the mind-brain conundrum quite well: "Men ought to know that from the brain and the brain only arise our pleasures, joys, laughter, and tears. Through it, in particular, we think, see, hear, and distinguish the ugly from the beautiful, the bad from the good, the pleasant from the unpleasant." Hippocrates and his followers seem to have grasped, as many of their contemporaries and successors did not, that the brain is the physical basis of subjectivity, which taken altogether we call the mind.

But how can basic physical material (even material as complicated as the brain) lead to abstract thought, emotions, and perceptions? Is it possible that each neuron in your brain has one specific function, so one neuron has the thought "Grandma" with a picture of your grandmother and a sense of how you are related? Can one neuron instruct you to keep your head down in your golf swing? Materialism runs into problems when we ask how the physical can give rise to the abstract.

The second stream of thought is dualism, formulated by René Descartes in the seventeenth century. He undertook an elaborate exploration of subjectivity in his *Meditations on First Philosophy,* in which he concluded that, unlike the external world, we have direct access to our own minds through introspection. By contrast, our knowledge of the external world is mediated by our sense organs, he argued. Descartes concluded that there must be mind-stuff, to which we have direct access, along with brain-stuff, to which we do not have direct access. The mind-stuff doesn't take up space, but the brain-stuff does. Hence, according to Cartesians, the universe is dualistic, as it can be divided into the mental and the physical, and so, the mind is altogether separate from the brain.

According to Descartes in the *Meditations,* to deliberately state "Cogito, ergo sum," "I think, therefore I am," is to identify an immediate and self-evident truth upon which all the rest of one's beliefs about the world can rest. He called the "cogito" an Archimedean point, a reference to Archimedes' observation that if he had an independent and stable place on which to stand, he could move the world. In an intellectual sense, Descartes' discovery of his own self-evident being met Archimedes' requirement. So the new world of the Enlightenment was to be one in which the earth was no longer the center of the universe. That central place was instead occupied by the thinking self.

Many see the subjectivism of Descartes' philosophy not only as a great victory of humanism that helped give rise to eras of inventiveness and industrialization (which it did), but also as the first step down a long road to narcissism and self-indulgence. But what I want to focus on is what might be considered the second most important assertion Descartes made in his *Meditations:* that there is nothing easier for me to know than my own mind. He reached this conclusion after convincing himself that it was always possible for his senses to be wrong about the "external world"—obviously the observations we make are often wrong—and that, by comparison, his knowledge of his own existence was rock solid. So, if he could never be fully sure he wasn't dreaming or making a mistake about the external world for some other reason, at least he could be sure he was thinking.

Based upon his own status as a "thinking thing," Descartes inferred the existence of the material world, including his own body. After all, he reasoned, though God might let Descartes' senses be fooled some of the time, God wouldn't let them be fooled all of the time, because that would be evil and it would be inconsistent with God's nature to be evil. This appeal to the nature of God without establishing on independent grounds that God does exist seems question-begging to many moderns, but the use of God's essential nature as what might appear to be the paradigmatic deus ex machina to tie up some philosophical loose ends is common in the history of philosophy.

Now, though, Descartes was left with another problem. His argument left him with two kinds of substance, one that doesn't take up space (his mind and the minds of others) and one that does (the physical world, including his body and those of others). How to get mind and body to coordinate in any particular individual? Descartes' answer was to call on what he knew of neuroanatomy. The only brain organ not duplicated on both sides and seated in the center of the hemispheres is the pineal gland, about the size of a pea just behind and a bit above the pituitary gland. The location and anatomic uniqueness of the pineal appealed to Descartes, who called it "the seat of the soul." Even now the gland's exact function isn't clear. Some believe it secretes a hormone that delays sexual maturity. Melatonin, dear to poor sleepers, is produced by the pineal gland. Descartes would have liked the claim made in 1976—but unsubstantiated by more recent methods of neurochemical measurement—that the gland produces tiny amounts of dimethyltryptamine (DMT), a psychedelic that might account for visual dreaming and mystical experiences.

However, no one seems to find Descartes' dragging in the pineal gland a satisfying solution to the metaphysical dualism his own argument generated. Descartes never explained how a material object such as this brain structure could be the site of the intersection of mind and body. Yet in spite of the limitations of his natural philosophy, Descartes is as responsible as anyone for turning the modern world's attention to the mysteries of mind.

Part of Descartes' mistake seems to have been the notion that we have direct access to our minds but not our brains. In spite of Descartes' conclusion that there is nothing easier for us to know than our own minds, we often do have quite a bit of trouble doing just that. In fact, Freud's psychology, another important product of the world created by the Enlightenment, is based on the view that we don't understand the mind well at all. So, for that matter, are the fields of psychology, psychiatry, and all the other disciplines that contribute to neuroscience. Contrary to Cartesianism, they find the mind to be the ultimate black box. From this standpoint, the idea that we could read the minds of others is laughable; modern "depth psychology" has been founded on the view that we can't even read our own minds without the help of psychotherapists, let alone those of other people. But now there is reason to believe that the black box is slowly, grudgingly, revealing its secrets.

Some of the most exciting work being done by neuroscience is made possible by machines that create images of brain activity, helping us also learn about our own minds at a much deeper level than we can learn through mere introspection. The whole field of neuroscience can be seen as one great challenge to Descartes and the dualistic philosophy he generated, yet we need to give Descartes his due as one of the precursors of neuroscience as well.

MAKING CONNECTIONS

Recent advances in neuroscience and cognitive philosophy may help us to make headway in answering the problems posed by materialism and dualism. A new theory called "connectionism" argues that the connections between physical substrates of the brain allow more complicated processes to function. Unlike pure materialism, this theory posits that the physical material itself doesn't give rise to "mind." And, to the chagrin of Descartes, connectionism argues that the mind must be a consequence of the brain's operations, not an independent substance. Instead, connectionism contends that it is the interaction (which is so complex that how these interactions produce mental operations is still quite mysterious) of physical parts that produces cognitive wholes.

Connectionism emerged from a second big problem in neuroscience that keeps coming up. Known as the localism-holism debate, the question is whether neurons and brain regions have specific functions or whether perhaps the entire system works together. The nineteenth-century phrenologists took localism to an extreme, mapping several dozen areas of highly particularized function that were said to actually bulge out of the skull. But others then showed that the destruction of regions of animal brains associated with certain functions could be compensated by other regions as the animal recovered. These observations seemed to support holism. For decades, though, the march of neuroscience seemed to be mainly in the direction of localism. Epileptic seizures were found to move from one part of the brain to another. It was shown that stroke patients who couldn't talk but did understand others had severe damage to the left frontal lobe. Later in the nineteenth century, when scientists applied electrical currents to brain regions, they got very specific muscular contractions.

In the twentieth century, the experimental results continued to seem inconsistent. For instance, K. S. Lashley wanted to know what part of rats' cerebral cortex stored their memory of a maze they had learned well. But when he cut different parts of the cortex, the rats were still able to get through the maze, leading him to formulate his theory of "equipotentiality" of different parts of the brain. On the other hand, the famous neurosurgeon Wilder Penfield found that when he activated certain groups of cells he would get very precise results, even getting patients to say particular words. The theory of connectionism was developed to resolve these apparently incompatible results. According to connectionism, lower-level motor functions, like muscle movements, are localized by higher-level operations, such as memory, due to connections between brain areas. Individual neurons do their work by being connected in systems to many other neurons. Even highly localized functions are really distributed over more cells than is obvious, according to connectionism.

DREADFUL FREEDOM

Although brain science may indeed progress without settling the mind-brain issue, it continues to present some perplexities for ethical, legal, and social policy. It doesn't seem plausible that the mind is totally independent of the brain, considering all the experiments and clinical evidence that show that brain injuries affect consciousness; but neither can we be wholly satisfied that the mind or mental activity is completely reducible to the brain. In his book *Consciousness Explained,* the philosopher Daniel C. Dennett elegantly explains the concern:

> If the concept of consciousness were to fall to science, what would happen to our sense of moral agency and free will? If conscious experience were reduced somehow to mere matter and motion, what would happen to our appreciation of love and pain, and dreams and joy? If conscious human beings were just animated material objects, how could anything we do to them be right and wrong? These are among the fears that fuel the resistance and distract the concentration of those who are confronted with attempts to explain consciousness.

But there are certainly those who don't share these fears because they don't see a problem. Consider this remark of Antonio Damasio in his book *Descartes' Error:* "The fact that acting in accord with an ethical principle requires the participation of simple circuitry in the brain core does not cheapen the ethical principle. The edifice of ethics does not collapse, morality is not threatened, and in a normal individual the will remains the will."

We might be satisfied with Damasio's observation from a philosophical point of view, that the moral life goes on as it does no matter what our theory of the mind-brain relationship happens to be. Of course, this idea could breed a morally hazardous form of moral materialism to the effect that no matter what we intend, the brain has a mind of its own.

But pragmatically, this cynical and defeatist interpretation only works in academic seminars, not in real life. When my child is mean to his friend on the playground, it would be irresponsible of me as a parent to shrug and explain to his parents that my son's behavior was merely the result of "matter and motion" (Isaac Newton's phrase for the radical reduction of everything to the physical). I can and should tell my son that what he did was wrong and that he is required to take responsibility, apologize, and control his behavior in the future. Children are taught that they are moral agents, that under normal circumstances they have free will, including the ability to choose between right and wrong.

But consider the results of certain experiments that might lead us to conclude that, in spite of Damasio's reassurance and my well-intentioned playground parenting, we have reason to worry about how free our will really is. Take the rather basic cognitive function of face recognition, associated with the fusiform gyrus, a structure in the cerebrum's left temporal lobe. It also seems to have a role in recognizing colors, words, and numbers. The left temporal lobe as a whole is a seat of comprehension, naming, and other linguistic functions.

There is lots of evidence that when the fusiform gyrus is damaged, people lose the ability to recognize faces. Those with damage to this area may have this distinct problem, but no difficulty with the more than two dozen other operations involved in visual comprehension. Using evidence from imaging machines, some neuroscientists claim that social judgments about trustworthiness seem to be based on the way certain faces look, and that these judgments involve a specific brain system. This perceptual processing, then, seems to be linked to social judgments that draw on the amygdala (an almond-shaped structure that is associated with emotions) and the prefrontal and somatosensory cortices (associated with planning, complex social behavior, and registering sensory information from the body surface). More disconcerting, other researchers have found evidence that some of the same areas are also involved in the preferential response to faces of one's own race. The obvious question is, as unprejudiced as we would like to think we are, what do these results tell us about how much free will we actually have?

But it is probably not a realistic concern that our every action is dictated by some neuronal action or blood flow in the brain. The idea that this might be the case is dubbed "reductionism," because every action or cognitive process is *reduced* to our brain activity. The philosopher Kenneth Schaffner distinguishes between "sweeping reductionism" and "creeping reductionism." Sweeping reductionism is the sort that some followers of Newton's mechanistic physics talk about: given a set of initial conditions in a system and the laws of the system, we can predict every subsequent state of the system. However, there's no reason to think that neuroscientific explanation is ever going to get that mechanistic. That certainly hasn't been the case for genetics, which instead has produced "creeping" or partial reductions that have lots of room at the margins. Very few diseases, for example, can be predicted based on DNA analysis alone. Many other variables are at play, including the kind of environmental toxins to which one is exposed.

Neuroscience seems to be in much the same position as genetics, indicating that brain states operate within a range of variability. (Of course, much of neuroscience involves genetics, so the way our genes code for our nervous system helps establish the framework of this variability.) If that's the case, then it's hard to make sense of the notion that there is always a chain of causes of our thoughts and actions wholly beyond our control.

Though sweeping reductionism will probably be a continuing background worry of philosophers, it won't have much place in the law. Many philosophers, beginning with Aristotle and all the way up to the modern British thinker H. L. A. Hart, have pointed out that there is a spectrum of actions from obviously unintentional to obviously intentional. In general, we don't hold people responsible for unintentional actions, such as when I fall in your lap while making my way past you to a seat in the movie theater. I apologize and, knowing it was an accident, you accept my apology. If I had looked at you first, shot you a grin and a wink, and then did a flying leap, you would be much less likely to excuse me. (We can't always plead accident to avoid some responsibility for our actions. As the English philosopher J. L. Austin observed, if I accidentally walk on your infant while she's sprawled on the living room floor you would rightly upbraid me and suggest I be more careful.)

Between the intentional and the unintentional are interesting and important cases of individuals who could have altered their behavior but chose not to. We often hold them at least partially responsible for their actions. Those are the ones who blithely chat with their friends while trying to make their way past me at the movie. In other middle cases, there are no clear opportunities to change behavior and outside influences that get in the way of change.

In these middle cases, it is important to understand the role that brain functioning has on our actions. Courts have begun to look at defense lawyers' arguments that certain medications compromised their clients' mens rea, the state of mind required for culpability. Sometimes, these arguments have been used when the defendant was on Prozac or similar medication. Typically, the court looks to experts to determine whether the drug could have led to the crime. Traumatic brain injury, especially involving damage to the prefrontal region, has been observed as leaving patients with the ability to reason about morality but not to act on the conclusion of that reasoning. If the damage can be shown through one of the new imaging technologies, this finding might be important to a legal decision.

BEWARE OF LOCALIZING

We need to be especially careful about how we talk about the brain to avoid the pitfall of trying to fit it and its functions into the language of psychology or ethics. Let's take the concept of aggression. Localism tempts us to think that there must be an "aggression region," while connectionism could lead us to believe that there is an "aggression network." Both ideas would be wrong. While clearly there is a lot going on in the brain that can teach us about the way aggression works and how to control it, it's a mistake to think that the social psychological construct called aggression corresponds to a region or a system in the brain.

New brain-imaging techniques seem at first to lend credence to the idea that brain processes can be localized to match up to psychological concepts. We need to be careful about overinterpreting the results of these studies. For example, the social psychologists Daniel Willingham and Elizabeth Dunn use the example of a brain-imaging study that asked people to make three judgments: whether an adjective described them, whether an adjective described a well-known person (e.g., the president of the United States), and a control task. The researchers found that different areas of the brain were activated when the subjects retrieved information about themselves than when they retrieved information about others. Does that mean that these areas are where the "self" resides? That would be an exciting finding, especially for followers of the eighteenth-century Scottish philosopher David Hume, who argued that the idea of the self was only one idea among countless others that self-conscious beings entertain. Has this neuroscience experiment nailed down the physical location of Hume's self idea?

But Dunn and Willingham note that there are at least a couple of reasons to be cautious about the notion that this study could tell us where the self idea is. First, the brain activation data might just be showing what happens when people attend to material about themselves. Second, if knowledge about oneself and others (knowledge about appearance, attitudes, etc.) is scattered over different areas, then those who know a lot about their own personality might store it in a certain cortical area because that's the area that is activated in this experiment, not because it houses the idea of self.

The problem is similar to using ordinary language to describe microscopic processes or events on a cosmic scale. As the logician Willard Quine observed, our language developed as a tool that works for our experience and interaction with middle-sized objects, a category that includes mountains and chairs and other creatures and even our planet and the visible stars, but not atoms or immense space. So we try to approximate these physical phenomena that are best described in mathematical terms with new words like "wavicles" or "quarks," or adapt analogical terms from ordinary language, like "string" theory. Similarly, the vocabulary of psychology wasn't intended to describe activities of the brain, especially not the brain as neuroscience is coming to understand it. So, we need to take care not to expect that a single area or even a single network is going to correspond to our psychological terms.

The best we can do, then, is look to more or less strong associations between our everyday concepts of thought and action that might be revealed by brain scans and other studies, though in practice these associations might be enough to guide interventions that could make a big difference in managing human behavior or even altering what brains can do. The assumption that this is so underlies the investment that national security agencies are making in neuroscience.

5:

BRAIN READING

PERHAPS IT'S FORTUNATE from a civil liberties standpoint that the brain isn't nearly as transparent, or as manageable, as might have been inferred from some of the early neuroscience or Descartes' philosophy. But depending on how permissive one is in the definition of mind reading, there are some pretty remarkable technologies in the works that will surely stoke paranoia. DARPA and other agencies have dipped their toes in this water with modest grants. Recall that only a small number of neurons in a system need to be read to identify a typical thought pattern. A *U.S. News and World Report* cover story for January 3, 2000, described some work at Lockheed Martin. The key concept is that, if neural activity could be detected at some distance from the brain surface, and if a computer had a dictionary of typical patterns associated with certain kinds of thoughts, then remote "brain prints" could be produced.

In an era of terrorism fears, the benefit of such a device is obvious: it could be placed in busy, sensitive public spaces such as airports. When individuals entered the terminal building with a pattern associated with certain violent thoughts, they could be taken aside for special screening. If the devices worked well enough, we could shed those annoying security lines and random searchers; only the very few with suspicious brain prints would be targeted for a close look. The Lockheed Martin researcher who is pioneering these remarkable studies claims that he can already deter-

mine when a subject is thinking about a certain number. If a single number is accessible to scanning, then so may be single letters, and, with a lot
more work, the strings of letters we call words, phrases, and sentences.

In 2002, there were press reports that NASA was working with an un-
identified private firm to develop a similar screening device. A *Chicago
Sun-Times* wire service story stated that a NASA aerospace research manager acknowledged that the agency was taking "baby steps" toward measuring airline passengers' brain waves. University of Maryland physicist
Robert Park told the paper that "we're close to the point where they can
tell to an extent what you're thinking about by which part of the brain
is activated, which is close to mind reading." But he also said it would
be very difficult to achieve such as result with a machine people walked
through.

Because of the long-standing stigma associated with mind reading as
pseudoscience, and the civil rights nightmare that would accompany any
such achievement, this is an area that can easily produce embarrassment.
Within three days of the *Sun-Times* story, NASA officials issued a statement in which they downplayed the project. "NASA does not have the
ability to read minds," the agency's director of its Strategy and Analysis
Division told the Federal News Service in what was, I suppose, intended
to be a reassuring statement, "nor are we suggesting that would be done."

MAPPING THE BRAIN

Philosophical discussions about mind reading could be rendered academic if certain DARPA projects are even modestly successful. Many
of these projects make use of functional magnetic resonance imaging
(fMRI), one of the most exciting windows into the black box. Magnetic
resonance imaging (MRI) is a method of visualizing anatomical details in
living things by means of magnetic charges. An fMRI scan takes advantage of the fact that when nerve cells are activated, their impulses metabo-
lize oxygen in the blood that surrounds the cells. The scan records the difference between oxygenated and nonoxygenated blood cells due to their
magnetic charges, so more active neurons can be distinguished from less
active ones. Combine this scanning ability with assigning an experimental subject specific tasks or experiences—the "functional" in fMRI—and

the result is a correlation between activated neural systems and mental activity.

A vigorous and lucrative MRI industry and other brain-imaging devices like PET and SPECT have vastly improved understanding of the brain's functional structures. Researchers have started to examine processes such as intending, speaking, and learning. Finally, a research industry has developed around the notion that brain function can be systematically linked to thoughts and actions by means of fMRI. Many of the results seem to concretize some of humanity's less admirable characteristics by linking them to brain processes.

For example, New York University neuroscientists led by Elizabeth Phelps have been able to correlate negative automatic evaluations of black faces by whites with activity in the amygdala, which processes emotion in the presence of stimuli. Stanford researchers then found that the face-recognition region was more active when viewing same-race faces. A string of studies supports the conclusion that different clusters of neurons are activated when different race-typical faces are viewed. Perhaps all such results tell us is that the evolutionary vestiges of "us" and "them" that were once adaptive but are now divisive are still very much hardwired, helping to explain why such determined social policy is required to counter deep-seated tribal tendencies. Other studies have applied fMRI to the way people make moral decisions, to how much they empathize with others, to levels of self-esteem as measured by frontal lobe activity, and even to how the brains of liberals and conservatives respond to campaign videos.

If the results of fMRI become reliable, this tool could present military officers with useful clues about a soldier's aptitudes for various assignments, supplanting old-fashioned personality inventories. In prisoner dilemma game experiments in which cooperation or competition is elicited, neuroscientists have found that when an individual chooses to cooperate with an opponent and then both win, circuits mediated by the chemical dopamine were activated, lighting up pleasure centers. Mutual cooperation seems to be a rewarding activity. Jobs or missions in which teamwork is crucial might best be filled by individuals whose dopamine-rich neurons are activated when put in situations that could elicit either cooperation or competition, as these people could especially enjoy working in a cooperative manner.

Another neurotransmitter, serotonin, is associated with feelings of well-being, and it seems to play a role in modulating reactions to stress. As a basic feature of combat, stress is unavoidable and stimulates both behaviors and the secretion of other chemicals that can be crucial for survival. But certain particularly sensitive missions, especially those done undercover, might best be performed by individuals who are big serotonin producers. Serotonin levels could be monitored during stressful situations, earning some selection for the most dangerous undercover operations in which keeping one's cool is more important than producing adrenaline. However, serotonin production is measured in the spinal fluid; it isn't clear what relationship those measures have to serotonin-mediated brain activity.

It's also not clear that imaging functional brain processes will be more reliable than old-fashioned methods for assessing cognitive performance. Traditional pen-and-paper surveys and behavioral observations might do just as well, and there aren't any good comparisons of fMRI and old-fashioned psychometric tests. But significant efforts are being made to see what fMRI can do for national security. The University of Pennsylvania's Institute for Strategic Threat Analysis and Response (ISTAR) is applying neural imaging to counterterrorism operations. Considering that fMRI can already detect simple lies, such as whether an individual recognizes a certain face, a suspect in a terror incident could be examined for truthfulness about certain acquaintances.

DARPA is trying to push brain-imaging technology further. A clutch of projects DARPA funded in its 2003 call on the possibilities of functional brain imaging. One contract awarded to a company in Eugene, Oregon, was for development of a "Head Access Laminar Optoelectric Neuroimaging System." The idea is to implant tiny sensors in a "head web" so that brain activity can be detected, transmitted, and reconstructed at a "Cognitive Workload Assessment" station. The dual use aspect of the device is especially attractive, as it can track the neural functioning of patients in the medical setting, even if they are ambulatory, and fits under the helmet of a soldier in combat.

A project from a Hawaiian firm, "Wireless Near-Infrared Devices for Neural Monitoring in Operational Environments," also involves the uses of new technologies such as wireless network chips and miniature lasers that permit "wirelessly monitoring neuronal activity." The abstract for this project summarizes in the painfully prolix lingo common to military contractors why several fields would be interested in such a device, including military research, medical care, and basic research.

> We propose to utilize these technologies to create a wireless whole brain functional brain imaging systems [*sic*] with applications in the cognitive neuroscience, brain-machine interface, and medical fields. . . . The proposed imaging system will yield a prototype that will demonstrate a mobile, continuous, non-invasive brain imaging system with spatial resolution on the order of centimeters and a temporal resolution able to image both the brain's immediate neuronal as well as its delayed hemodynamic response. This combination of characteristics will make this prototype a state of the art imaging system with far reaching applications and markets extending to: Military Research laboratories currently involved in brain imaging, Laboratories that in the past could not afford the current state of the art brain imaging technology, and Hospital patient monitoring. . . . We envision transitioning the prototype into several products. One of these will be directed towards continuous monitoring in restricted environments such as the cockpit of flight simulators. Another product will be catered to the general brain imaging research community where an emphasis will be placed on easy extensibility by the community so that the product can continue to mature through industry-academic collaboration. Still another area of interest for product development is in the use of such a system for monitoring of blood pooling wounds, particularly hematoma development but also, with some changes in geometry, in other injuries occurring throughout the body (such as internal bleeding).

English translation: They're going to combine into a single system techniques for measuring brain activity and wirelessly transmit all that information to a computer that will interpret the information for various purposes.

A second project description in this category ("Wireless Near-Infrared Devices for Neural Monitoring in Operational Environments") notes that "a wireless monitoring device that offers both neuronal and vascular signals has a huge commercial potential." The list of applications includes brain research that can be done with portable monitors, including "the study of normal brain development in infants, the diagnosis and follow-up of cerebrovascular diseases, and psychiatric syndromes in adults and children." And a third project offers a similar laundry list of applications:

The market for non-intrusive portable monitoring by means of non-invasive brain monitoring offers a most exciting and significant break-through, impacting many industries. Early adapters are expected from the military for training under stress; medical—head trauma evaluations; educational—diagnosis of learning disabilities; and law enforcement—for interrogation. Medical research will also benefit from this research and development effort, because many brain studies may be improved by portable monitoring of functional activities including stroke rehabilitation and epilepsy. Other research benefits are for the study of normal brain development in infants, the diagnosis and follow-up of cerebrovascular diseases, and psychiatric syndromes in adults and children.

These projects are closely related to the Defense Department's augmented cognition effort, represented in the observation in one abstract that it will "use non-invasive physiological monitoring to aid in managing the workload of military personnel in a multi-task or high stress environment. Numerous physiological sensors are currently available for monitoring physiological parameters such as electrical activity in the brain, heart rate variability, respiratory rate, vascular blood volume, and skin conductance, all of which can be used to evaluate stress and cognitive workload." For a study called "Personnel Monitoring for Assessment and Management of Cognitive Workload," the abstract states:

> Assessing the levels of stress and cognitive workload of numerous personnel allows for work to be delegated efficiently to those operators who are the most physically and mentally equipped to carry out a given mission or responsibility. Potential natural dual-use applications for a robust and reliable personnel assessment tool include use by commercial pilots/aviators, fire/rescue personnel, police, and others operating in high stress environments. Additional dual-use applications for this technology include use by pharmaceutical companies to assess the efficacy of stress related drugs, and use as an early detection platform for diseases such as Alzheimer's Disease and Parkinson's Disease.

Other studies propose to develop software to support automated workstations "for the monitoring of the operator's cognitive state;" technology that "would analyze brainwave patterns associated with higher order executive processing" and other measures to identify cognitive workload and stress; and a "'neurogenetic' agent framework . . . to model, simulate, evaluate and compare the leading architectures for intelligent agents."

All these projects involve monitoring the nervous system at a gross level. Because they are, after all, pitches to government agencies by private contractors, they might not be able to deliver, especially to the extent that they are based on variations of electroencephalograms (EEGs). Typically, EEGs are good at picking up brain activity in real time, but not very good at identifying where in the brain the signal originates. These efforts also fall short of the most literal notions of mind reading, though they might produce substantial new knowledge about the brain in all sorts of challenging situations, information that can improve the quality of life for many who are ill and help avoid medical crises. Spinoffs could lead in directions that are not specifically national security oriented, but it is well within DARPA's mission to take as long as needed to find out.

THE "BRAIN FINGERPRINTER"

A slew of studies and at least one commercial product are aimed at using neural activity in lie detection. Penn psychiatrist Daniel Langleben is using fMRI to identify brain regions associated with lying. He and his team have concluded that "cognitive differences between deception and truth have neural correlates detectable by fMRI," with increased activity in the "anterior cingulate cortex (ACC), the superior frontal gyrus (SFG), and the left premotor, motor, and anterior parietal cortex [sic] . . . specifically associated with [deception]." Happily, it's beginning to look as if we are wired to tell the truth. A review of studies on the use of fMRI to detect lying reports that attempts to deceive are associated with activation of executive function centers, especially the prefrontal and anterior cingulate cortices, but truthful responses don't activate any particular areas more than others. "Hence," a University of Sheffield neuroscience group concludes, "truthful responding may comprise a relative 'baseline' in human cognition and communication."

Techniques aren't yet specific enough to predict when a particular person is being intentionally deceptive. However, there are some indications that refinement of fMRI for lie detection is possible, based on the fact that our natural inclination to be truthful forces the brain to work harder when we lie. Harvard's Giorgio Ganis and Stephen Kosslyn have found that well-organized lies involve activation of many parts of the brain—a convincing lie requires concentration—and rehearsed lies can be distinguished from spontaneous ones. Similarly, a Medical University of South Carolina team found increased activity among lying young males in the anterior cingulate and the orbitofrontal cortex. The forensic and national security implications of a reliable individual brain scan for lie detection are obvious.

Then there is the "brain fingerprinter," an early attempt to catch the wave of neuroimaging for lie detection. Among neuroscientists, the brain fingerprinter is notorious for the claims that have been made by its promoters. The concept behind the main product of Brain Fingerprinting Laboratories, Inc., is to identify memory traces through brain wave responses rather than fMRI. The basic idea is that the brain reacts spontaneously to stimuli it recognizes and this recognition can be recorded as a single rate of oscillation. If a subject's brain reacts a certain way to, say, a picture of a crime scene or a terrorist training camp but the subject denies ever having seen what is in the picture, a lie supposedly is detected.

Developed by the inventor Lawrence Farwell, (the general approach, called computerized knowledge assessment, or CKA, was originally reported by a team of neuroscientists in 1988) the system takes advantage of electrical responses in a set of neurons to stimuli such as words or images known as event-related potentials (ERP). In particular, brain fingerprinting takes advantage of a P300, an oscillation or bump in the line traced by electrical detection equipment that occurs three hundred milliseconds after the stimulus, before the subjects are aware of it and therefore before they can change it. The P300 is part of a larger electrical response in the brain, the memory and encoding related multifaceted electroencephalographic response (MERMER), that, once combined in an algorithm with the P300, is claimed by the company to have passed muster with a definitive determination of truth-telling in more than two hundred subjects.

In spite of the general agreement among neuroscientists that brain imaging is far too unreliable to carry much weight as evidence and may never get to that point, the company claims to have achieved recognition in the legal system. That would set the brain fingerprinter apart from the old-fashioned lie-detector test, which isn't admissible in court in any state except New Mexico. On its Web site, the company seems to take credit for the release of an Iowa man, Terry Harrington, from prison twenty-four years after a murder conviction. The case against Harrington was based

largely on the statement of a single witness who later said that he was trying to avoid prosecution for the crime. There is debate, however, about
how much of a role the brain fingerprinter had in the case. An Iowa dis-
trict court ruled that the P300 test results would be admissible but rejected Harrington's petition on other grounds, saying that the P300 results
would not have influenced the outcome anyway. Harrington's lawyers
won a retrial order from the Iowa Supreme Court, but instead of holding
another trial the state released him in October 2003.

With this much wind at its sails (or, as the critics would have it, hot
air), brain fingerprinting has been of special interest to law enforcement
agencies since 9/11. P300 does not seem to have been admitted to any other court proceedings, but it has been used by criminal investigators in
other cases. The company Web site uses dramatic language to promote its
possible benefits. "In a terrorist act, evidence such as fingerprints or DNA
may not be available, but the brain of the perpetrator is always there—
planning, executing, and recording the crime." Records encoded in neurons could "help identify trained terrorists before they strike, including
those that are in long-term 'sleeper' cells," the company claims. Screening visa applicants and those who have access to classified information
are also listed as amenable to P300 detection. And, like so many other
examples of brain-related technology, the company further suggests that
systems it is developing can help in early identification of Alzheimer's disease and cognitive decline. Commercial possibilities include testing to see
if advertising is effective in registering as a memory and in investigations
of such nonviolent crimes as insurance fraud.

Although the jury of science is still very much out on the ultimate
power of the P300 bump detector, it or similar technologies could still
gain a foothold in the legal system. As Stanford law professor Hank Greely told *U.S. News & World Report,* "As long as a judge is convinced that
something is basically scientifically reliable, she can let it in."

Helped by its jazzy if oxymoronic and misleading name and bolstered
by the desire of law enforcement agencies to get any help they can in their
antiterror efforts (the CIA and FBI invested one million dollars in basic
research contracts, though it's not clear how much they are actually using
the technology), brain fingerprinting is off to a strong start in the federal
funding universe. Oregon Senator Ron Wyden and California Congress-

man Michael Honda have argued for using computerized knowledge assessment in the war on terror. Silicon Valley entrepreneur Steven Kirsch has suggested that the combination of a CKA test and an iris scan would create a security profile for air travelers.

Such a rush to judgment would be unfortunate. As a forensic device, critics of brain fingerprinting note that the test really measures whether the subject is familiar or unfamiliar with a crime scene, not deception. In the only published study on the subject, the ERP detector didn't do much better than guesswork. Also, if the subject has forgotten an event or if he or she is mentally ill, the response could be altered. There's also evidence that people can produce P300 waves by deliberately thinking about stimuli that have never taken place except in their imagination, such as a group of students in one study who produced a bump after being told to think of their instructor slapping them. Terrorists' brains might show that they're well aware of terror targets and methods, but so might the brains of journalists and intelligence experts—and teenage boys who find jihadist Web sites titillating.

One stumbling block to validating the P300 is the fact that the techniques used to analyze brain fingerprinting results are proprietary, so they can't be analyzed by independent scientists. University of California, San Francisco, deception researcher Paul Ekman is making a start at an alternate, low-tech approach. With NIH sponsorship and decades of experience, Ekman has reported that trained observers using behavioral clues—no brain scans—can identify lies about 80 percent of the time. With DARPA funding he is investigating whether highly motivated liars such as terrorists might be more effective at evading detection.

Procedures like Ekman's might do just as well as fMRI approaches, but that obviously isn't stopping efforts to commercialize imaging for lie detection. In 2006, two companies, No Lie MRI and Cephos, announced plans to make their technology available. Both are advised by highly respected neuroscientists who have been quoted as claiming that they can tell when someone is lying approximately 90 percent of the time. Civil libertarians and neuroethicists have called for restraint. One suggested approach: government regulation of MRI for lie detection until its safety and efficacy have been proven.

MIND READING REDUX

Though the claims for brain fingerprinting might turn out to be exaggerated, its surface appeal has a lot to do with our fascination and anxiety about the idea that someone else can know what we're thinking, maybe even better than we can. The twentieth-century idea of "mind reading" came both from an awareness of how hard it is to know the mind and from the excitement that science seemed finally capable of resolving the mystery of mind. On the dark side, the idea of mind control was an implicit acknowledgement that these breakthroughs, when they came, would bring new danger as well.

For years, I have corresponded with several very bright and highly functional people who are absolutely sure that at some time or another they have been the victim of mind control experiments by a government agency. Once I asked one of them if anything would ever alter her view about this; she acknowledged that probably nothing would, such is her certainty about her victimization by surreptitious forces. My own experience with government—on the staffs of presidential advisory committees, in congressional testimony, and so forth—makes me doubt that such experiments could be kept quiet for decades. Our government just isn't that airtight. So, I'm no conspiracy theorist.

Yet, considering all the attention that DARPA and other agencies are giving to speculative attempts to monitor and augment brain function with technologies such as fMRI and the brain fingerprinter, there is a basis for concern about whether we are going down a scary road, one that would justify the paranoia of the mind control cult. In fact, enormous technical obstacles lie in the path of anything that could pass muster as a reasonable sense of mind reading. Perhaps the biggest obstacle is the way that the brain encodes information. In recent decades, the belief that neurons were dedicated to specific activities, such as hearing a certain pitch or moving a certain limb, has been undermined. It turns out that neural cells change their duties throughout our lives. Ironically, that is just the theory that one of the grandfathers of neuroscience, William James, offered in his landmark text, *The Principles of Psychology,* in 1890, when he referred to the marvelous plasticity of the neural material.

Another recent deviation from previous neuroscience orthodoxy also undermines the literal idea of mind reading. The Briton Lord Edgar Adrian showed in experiments with frogs and eels that the more intense a stimulus is, the more a neuron fires, as much as two hundred times a second. The firing rate, or "rate code," was subsequently found to be the way the brain sends information. So along with the supposed specificity of neuronal functioning, it seemed that zapping a certain set of neurons with a certain amount of electricity would get the brain to do what you wanted it to do. Conversely, a "dictionary" of rate codes and neurons could tell us what the brain is thinking at any particular time. But, again, the picture was far too simple. Some variations in firing may be meaningful, but others may not be. The simple fact that one rate code is bigger than another doesn't mean that more information is being passed around.

So even though some sort of "dictionary" of brain events is not out of the question (and neuroscientists disagree about how plausible such a dictionary is), the obstacles are formidable. When electrodes are hooked up to rats' brains, the firing patterns differ each time they run a maze. As Berkeley's Walter J. Freeman told *Discover* magazine, even if we could solve the mystery of neural coding and build an intelligent machine on that model, "we won't be able to read its mind either." If Freeman is right, we may be caught in an infinite regress that marks the limits of insight into the workings of highly evolved minds.

Dartmouth professor Adina Roskies makes a different and equally crucial point about the limits of any supposed dictionary of the mind. The philosopher and neuroscientist points out in an article in *Cerebrum* magazine that understanding the way the brain works through neural imaging does not imply an ability to manipulate it: "This worry, I think, stems from a misunderstanding about the sort of information that brain imaging yields." Roskies continues:

> Neuroscientific knowledge is descriptive: it allows us to correlate brain activity with behavior and enables us to figure out what sorts of brain activity are involved in various types of cognitive tasks. Descriptive knowledge, if fine-grained enough, may also be predictive. It may enable us to anticipate that a person exhibiting a certain pattern of brain activity will act in a particular way, or will engage in a particular kind of cognitive processing. But the ability to predict is not the same as the ability to control. Using neuroimaging to aid in understanding brain-behavior relationships does not allow us to control

behavior any more than taking picture at a busy intersection enables to control the flow of traffic.

Roskies' traffic analogy is apt. Of course, knowing something about traffic flow can tell urban planners something about how to influence it. Creating left-turn-only lanes or prohibiting left turns during rush hours are familiar examples. Similarly, Roskies speculates that a company might decide to pitch advertising based on what it has learned from brain scans about responses in brain areas associated with pleasure, and perhaps even correlate that with activity that relates to the impulse to buy. (Speaking for all the world's authors, I wish I knew how to do that.) But again, she notes, all advertising is an attempt to manipulate behavior and, indirectly, the brain. Brain scan data would just provide an alternative to market research and might not be as reliable. And, of course, even the most effective advertising can be resisted after one has some practice in the marketplace. After all, we don't roll down the aisle tossing everything into our cart that strikes our fancy, much as we might want to. We have devised systems for interfering with the desire to scratch the itch created by the most ingenious marketing—at least most of the time. But is that only because the science hasn't progressed far enough to be really invasive?

As we keep our minds open (so to speak) to the technical possibilities, we shouldn't be dazzled by fancy mind reading claims for exotic machines. As Dartmouth neuroscientist Michael Gazzaniga puts it in his book *The Ethical Brain:*

> Neuroscience does not yet have incontrovertible evidence of how thoughts are represented in EEG scalp recordings, let alone in the brain, and it may well be that while all thought is generated in the brain, we may never be able to read those thoughts. . . . "Mind reading" technologies do not, in fact, read the mind. They are just another set of data to be interpreted contextually. Neuroscience reads brains, not minds. The mind, while completely enabled by the brain, is a totally different beast.

There is by no means universal agreement among neuroscientists on this point. Penn's fMRI expert Daniel Langleben has said that "in the long term, I think we will have technologies powerful enough to understand what people are thinking in ways that are unimaginable now. I think in 50 years we will have a way to essentially read minds."

DETERMINED TO BE INDETERMINED

Let us set aside questions of literal mind reading and take up the somewhat more straightforward relationship between predicting behavior and controlling behavior. This question is of course critically important for our ability to assess the social and military implications of modern neuroscience. New York University professor Paul Glimcher has pioneered neuroeconomics, the field in which neuroscience and economic theory meet. His physiological studies of rhesus monkeys are gradually enabling him and his team to predict what decisions the monkey will make under certain circumstances based on neural activity. Glimcher told me that he believes that as this kind of work progresses, it will be possible to predict much decision-making behavior to a very high degree of accuracy, even when the subjects try to mask their intentions. Even now, once he has the opportunity to examine human subjects through fMRI and learn their patterns, he can identify thoughts about a simple object in spite of attempts to block the image. ("Don't think about elephants.") But, like Roskies and others, Glimcher does not think that perfect accuracy will ever be attainable, even with far more sophisticated algorithms and instruments than we have now.

Glimcher's reasoning takes us back to the philosophical question about free will. Like other neuroscientists and philosophers, he argues that even though in a causal sense all our thoughts and actions may be determined, it does not follow that they are all predictable. One way of thinking about this is that evolution has produced a creature (humans and perhaps other primates) that has the ability to deliberately and nondeterministically enter into the "chain of causes" that leads to thought and behavior. Think of us as carbon-based embodiments of quantum physics. We are in effect deterministically indeterminate systems, in Glimcher's view.

Yet Glimcher is also confident that some neuroscience-based applications will be useful in security situations, perhaps being able to predict to a rate of 90 percent or better what an individual will say or do. But again, that doesn't imply that others will have the ability to control someone's thoughts and actions. Even for mere prediction, Glimcher also agrees that psychologists using low-tech methods might do just as well as neuroscientists working with their high-tech and seemingly more "invasive" equipment. But the temptation to apply the more "hard science" innovations may prove too great to avoid, as they often provide the illusion of greater predictive power than someone armed only with a clipboard.

COGNITIVE LIBERTY

If mind reading and ideas about eventual mind control are oversold, and whether some of the technologies I've described turn out to be successful or only a passing hyped fantasy, it's important to remember that the basic point of neuroscience is to gain knowledge about the human that might someday be important in fighting neurologic diseases. Yet not even the most seemingly benign brain research project—even one that is funded only by civilian health research agencies—can totally escape the attention of those concerned about civil liberties.

Neuroinformatics is the name given the combination of neuroscience and information scientists (computer scientists, engineers, physicists, and mathematicians). Perhaps the most prominent comprehensive effort in neuroinformatics has been the National Institute of Mental Health's Human Brain Project. Funding for the HBP, which has been around at least on paper since 1993, has at best an uncertain future. From a scientific standpoint, the inadequate support of the HBP is unfortunate, for the idea is intriguing: to "map" the brain, creating a National Neuronal Circuitry Database. The HBP's primary goal is to analyze functional interactions among neurons. In the longer term, the HBP is also intended to "make available to researchers powerful models of neural functions, and facilitate hypothesis formulation and electronic collaboration," says the NIMH. Numerous universities in the United States and abroad now participate in the HBP with funding from the NIMH and other NIH institutes. Examples of their work include applying MRI technology and new computational techniques to the developing brain, integrating functional models of language use with the structural framework of the cortex, and modeling neural systems such as the olfactory (sense of smell) pathway.

Even if the HBP effort falters, the concept it represents is both admirable and socially challenging. Wrye Sententia, director of the Center for Cognitive Liberty and Ethics, acknowledges the HBP's importance in advancing knowledge of the brain to develop treatment for many terrible diseases. Writing in the *Journal of Cognitive Liberties,* she sees it as part of a bigger and disconcerting picture:

> "While the idea of a mapped brain may seem ludicrous given our brains' operative complexity and our limited understanding of how the brain works," Sententia writes, "the fact that a world-wide team of HBP scientists and specialists are, or shortly will be, working to track and record composite patterns of thought is indicative of the trend towards brain monitoring. If guidelines and patterns can, as proponents of the HBP hope, be overlain to "map" typical brain features and attendant thought functions, these guidelines could mean less cognitive liberty and more mental street signs, more data-mind surveillance.

Interrogations would seem to present obvious potential applications of brain maps along with fMRI and related brain-imaging technologies, but their use would run up against significant legal obstacles. A subject volunteering to answer questions is acceptable, but if the technology is used to obtain or confirm information that the subject has not agreed to disclose, that would raise questions. Unlike traditional lie detectors that cannot prevent individuals from altering their emotional responses to their own statements and thus evading detection, an fMRI scan measures blood flow as an indication of a thought process. Activity in a certain neural pathway cannot be deliberately controlled by a subject; thus, nonvoluntary disclosure is possible. In this respect, even physical coercion could be less invasive (although more frightening and injurious) than a valid fMRI scan.

Sententia and other cognitive liberty advocates often seem to appeal to a slippery slope argument, cautioning that although current intentions might be sound, the direction in which they lead is ominous. Other than close monitoring and perhaps regulation, it's not clear what remedy there is for the prospect of this alleged slide toward a radical loss of privacy and perhaps even individuality. Nor, as Sententia and others note, is it at all clear that human neurobiology will ever be so well understood as to permit the worst-case scenarios to unfold.

FEED YOUR HEAD

It's especially hard to assess the plausibility that something such as mind reading or mind control is feasible through the kinds of devices I've described in this chapter. Many of the technologies do seem hyped; just because national security agencies are spending money on them doesn't mean they are a sure thing, but that's often enough to make conspiracy theorists feel vindicated. With brain theory as inconclusive as it is, there are bound to be conflicting claims among neuroscientists about what's technically possible and what isn't. Since neuroscience hasn't come close to finding the boundaries of its possibilities yet, that uncertainty is likely to persist for a long time.

Unfortunately, as we have seen in so many other cases, expert disagreement about whether a technology or product line delivers what it promises doesn't mean that it won't be prematurely adopted. This has happened in the pharmaceutical industry on more than one occasion, often with tragic consequences. Besides the possibility of excesses in the legal system produced by judges who get too excited by a cool device, there are also the temptations for insurers or employers to require brain scans to assess risks and aptitudes, or educators and social workers to evaluate an adolescent's propensity to violence or other antisocial behavior, not to mention commercial scanning services already open for business in some places. Regulating the introduction of devices spun off from neuroscience into various social institutions is going to be one of the big social policy challenges of this century. With military and intelligence needs on the cutting edge of these developments, the policy challenges are going to be still more daunting.

6:

BUILDING BETTER SOLDIERS

THE HUMAN BEING IS THE OLDEST INSTRUMENT OF WARFARE and also its weakest link. Although astonishing and terrifying "improvements" have been made in the devices of conflict over the millennia, soldiers are still basically the same. They must eat, sleep, detect danger, discern friend from foe, heal when wounded, and so forth. The first state (or nonstate actor) able to build better soldiers using medical enhancement technologies will have taken an enormous leap in the arms race. The concept of "an army of one" and the recent shift from soldier to "warfighter" in the military lexicon (as University of Pennsylvania bioethicist Paul Root Wolpe pointed out to me) are tied into the goal of building a more self-sufficient individual warrior. However better soldiers are built—and there's good reason to believe that the warfighter of the late twenty-first century will be enhanced—the fighter's brain will have been the object of greatest interest.

PERCHANCE NOT TO SLEEP

Fatigue-induced error is already being targeted, as death by "friendly fire" is a shockingly frequent occurrence that can partly be attributed to the chaos and confusion of combat but also to the sleep deprivation that accompanies lengthy engagements. For example, two American pilots accidentally killed four Canadian soldiers and injured eight others

114

in Afghanistan in January 2003. This was a horrifying international in-cident that resulted in the courts-martial of the pilots before the charges were dropped. The tragedy provided an unexpected glimpse into the Air Force's interest in sleep. Unnoticed by many, the pilots' attorneys pointed out that their clients had been taking Dexedrine, sometimes called the "go pill" in the military, otherwise known as "speed." It was alleged that am-phetamines like Dexedrine are commonly prescribed to keep pilots alert for thirty-hour missions, though questions have been raised about their safety. Their use can also lead to drug dependence.

So the U.S. Air Force is considering alternatives to amphetamines, es-pecially a medication that has also gotten the attention of long-distance business travelers who cross time zones: modafinil. Approved by the FDA in 1998 and marketed as Provigil, modafinil is used to treat narco-lepsy and to help with the sleep disorders that come with diseases such as Parkinson's, Alzheimer's, and multiple sclerosis. Modafinil, it should be emphasized, is not a stimulant as we've come to understand the term. Rather than bombarding various parts of the brain with arousal signals, modafinil apparently nudges the brain toward wakefulness through spe-cific pathways, perhaps by increasing serotonin levels in the brain stem. The precise mechanism is still not well understood.

The temptation for healthy people to use it as an antisleep agent is tre-mendous; some report that a dose leaves them as refreshed as a short nap. Frequent fliers are already getting prescriptions for the stuff, and it's sure to be the next craze on college campuses among students who want to pull all-nighters or just be able to party hearty for days. Some health educators worry that modafinil will rival or even replace stimulants such as Ritalin, which in turn replaced amphetamines such as Dexedrine that were the "uppers" of choice when I was in college in the early 1970s. Long-distance truck drivers, who too often grab just a few hours of sleep to stay ahead of schedule, are also obvious candidates for use and, perhaps, abuse. Misus-ing modafinil wouldn't take the familiar form of achieving a "high," but taking it for inappropriate purposes or to an extreme that has not been shown to be risk free.

Another candidate population for the drug is the large number of workers who shift from day to night schedules and back again. They often complain of drowsiness during the work period and insomnia when they want to sleep. The Air Force Office of Scientific Research and a company called Cephalon sponsored a study by Harvard and Penn researchers in which sixteen healthy subjects were deprived of sleep for twenty-eight hours, like shift workers, and then obliged to sleep from 11 a.m. to 7 p.m. for four days and to stay awake at night. The subjects on modafinil did far better on cognitive tests than those on a sugar pill.

A few news outlets made unconfirmed claims that American soldiers were using modafinil on the way to Baghdad in 2003. That wouldn't be surprising. A solution to sleep has been a minor-league Holy Grail for war planners since time immemorial. Guards at China's Great Wall chewed an herb containing ephedrine; Incan fighters munched on coca leaves; nineteenth-century Bavarian officers gave their men cocaine; several countries' soldiers used amphetamines during World War II; and, of course, armies consume countless tons of caffeine and nicotine. Even modafinil has been around for decades, used by the French Foreign Legion in World War I and, ironically, as a treatment for cocaine addicts. Like so many other compounds, it was taken off the shelf as its other potential properties became apparent. French soldiers took modafinil in the first Gulf War, and the *Guardian* newspaper reported in 2004 that the UK Ministry of Defence had bought 24,000 tablets of the drug.

But is modafinil truly a wonder drug, able to increase both wakefulness and vigilance while amphetamines often cause their users to be anxious and nervous? Double-blinded, placebo-controlled studies have already shown the antisleep properties of modafinil, with some able to stay awake for more than ninety hours. More complicated is the question of when alertness begins to fade. Does the drug mask natural sleep needs but fall short in keeping people as functional as they think they are? This could be critical for pilots and soldiers, who should not overestimate their readiness. In the longer term, the endocrine and immune systems may be compromised by lack of sleep.

SLEEPLESS SOLDIERS

Military scientists are working on the safety questions. One researcher at the Air Force's Brooks City-Base in San Antonio told *Pentagram*, an online newsletter, "We are trying to find out if this is better than what we have or just another drug to help pilots stay awake. It's too early to say if it's a better choice right now. All indications say Modafinil is a safer drug, but we don't know that for sure. That's why we go to exhaustive measures to make sure they're [*sic*] safe." Safe in terms of sound judgment in combat, perhaps, but what about the effects down the road for people who have been on and off the drug for years?

The precise function of sleep and the long-term risks associated with sleep deprivation aren't well understood. There is evidence that during sleep, memory and learning are consolidated, and that the brain refreshes its store of energy then. Studies have shown that people who sleep only four hours a day for an extended period show an increase in insulin resistance, a prediabetic symptom. But without a proven explanation for the purpose of sleep, it's hard to assess the downside of doing without, other than the obvious discomfort that nonsleepers experience, the attendant loss of concentration, and the increased accident risks.

An intervention that could minimize sleep while retaining cognitive capacity would be a significant advantage for a military force. Infantrymen commonly subsist on three or four hours of sleep nightly for weeks at a time. Special Forces personnel may be awake for several days in search and rescue operations. As a way to squeeze out more productivity from soldiers, minimizing sleep has long been at the top of the wish list, hence DARPA's Preventing Sleep Deprivation (PSD) program. One hundred million dollars in grants is being spent for research on "prevention of degradation of cognitive performance due to sleep deprivation." As the PSD announcement put it, "As combat systems become more and more sophisticated and reliable, the major limiting factor for operational dominance in a conflict is the warfighter. Eliminating the need for sleep while maintaining the high level of both cognitive and physical performance of the individual will create a fundamental change in warfighting and force employment." DARPA's Defense Sciences Office described the problem as part of the agency's Continuous Assisted Performance program:

> Continued assisted performance really asks a basic question. Can you prevent the cognitive deficits that occur in sleep deprivation from occurring? If you can prevent bad decisions from being made during sleep deprivation, you can dominate the battlefield by limiting the requirement for sleep. If you cannot prevent those changes from occurring, can you reverse them when they

have occurred? Or can you create alternate pathways and expand the available memory space, so that people can retain cognitive function under tremendous stress and sleep deprivation?

The PSD effort includes investigations of another class of drugs, the ampakines, which show some promise in treating dementia and symptoms of schizophrenia by improving cognition when used with antipsychotic medication. Unfortunately, so far clinical trials have not found therapeutic value for these drugs. However, results from a biotech company–sponsored study at Wake Forest University using an ampakine drug in sleep-deprived rhesus monkeys were encouraging. The monkeys' performance was reduced 15 to 25 percent when sleep deprived, and reaction times doubled. But when these monkeys got a single dose of Ampakine CX717, their performance deficit was eliminated, as were sleep deprivation changes in their EEG. An unpublished human trial sponsored by the company that makes CX717 reported that sixteen men deprived of a night's sleep did better on memory and attention tests after taking the drug. The scientist who conducted the study said, "We didn't see any adverse events."

Columbia University neuroscientists using fMRI have found that some neural pathways work better than others under sleep-deprived conditions, leading to speculation that it may be possible to train people to use these pathways. Such an approach might avoid the use of drugs. Transcranial magnetic stimulation (TMS) has been used by neurologists for years to make sure that a certain neural circuit is still intact. Using a hand-held magnetic coil on the scalp, a very precise current can be passed into the brain, apparently without injury, though a seizure can be triggered if correct practices aren't followed. This seemingly low-risk approach can help both the operator and the subject know when certain pathways are in use, a kind of biofeedback loop.

Modafinil and these other features of the Preventing Sleep Deprivation program are likely only the beginning of intense efforts to control sleep-wake cycles, driven by a multibillion-dollar demand among those who want to sleep only when they want to sleep. *Wired* magazine reported in 2003 that a company called Hypnion, based in Worcester, Massachusetts, is attempting to develop drugs even more effective than modafinil. Hypnion scientists put radio transmitters on animals that have been given experimental drugs to record various functions, including sleep, using a system developed with Air Force and Defense Department funding. In the company's labs, transmitters on mice and rats send radio signals to computers that record the drugs' effects on biological processes, particularly the rodents' sleep patterns.

The neuroscientific key to sleep lies in a part of the hypothalamus called the suprachiasmatic nucleus (SCN), the brain's biological clock that was discovered in 1972. About the size of a pinhead and nestled deep within the brain, the SCN with its roughly twenty thousand neurons is the pacemaker for circadian rhythms in mammals. Scientists have discovered in animal experiments that if the SCN is cut or removed, the sleep-wake cycle can be profoundly disturbed. The SCN's normal twenty-four-hour clock manifests the intimate connection evolution has given us to our terrestrial home. But when people are deprived of light, the SCN runs on a twenty-five-hour clock, so for some reason that is our innate length of a single day, a fact that helps explain insomnia and other sleep disorders in people who are blind. Though subject to some variation, the clock can be reset by exposure to light signals transmitted from the retina or by other time cues, such as a meal at an odd hour.

Scientists who are not directly involved in the PSD program or its funding seem moderately hopeful that the research can bear fruit. The distinguished Harvard physician and neurobiologist Jerome Groopman has written that

> the widespread assumption that sleep is necessary was supported by early studies of sleep-deprived rats: they suffered deterioration not only in behavior but in body metabolism and immune defenses. As repeated experiments have verified, when rodents are prevented from sleeping they often die of sepsis, with some succumbing after only five days, the hardiest lasting a full month. Yet such effects have not been seen in human subjects. And, surprisingly, there is very little hard data showing that prolonged sleep deprivation truly has deleterious effects on us. The lore that it can cause psychosis dates to the Korean War, when Chinese Communists were said to torture prisoners by preventing them from sleeping; however, later researchers have concluded that the psychosis resulted from the kinds of stimuli and stresses applied by torturers under these gruesome conditions, rather than from the lack of sleep per se.

And Groopman quotes Penn sleep researcher David Dinges, who raises a provocative question about Boeing's plans for a passenger jetliner that will fly around the Earth and need to land only once in twenty hours. "How should the crew sleep, if at all? What are the rules that apply to sustain work on flights like that?" This problem is only one of a number that confront a culture that has been trying to reduce its dependence on sleep since the Industrial Revolution, as Dinges views it. "Now is the time to have an open and frank discussion on how far we will go as a culture, what are our priorities, how regularly do we want to manipulate our brain chemistry? What are the limits?"

Whatever the limits are or should be for the civilian world, security issues may be seen as superseding them. In fact, nonhuman mammals may hold the key that will change evolutionary sleep patterns once and for all, with the military as the leading edge. Dolphins seem to keep parts of their brains awake all the time to control their breathing and come up to the surface while other brain regions are allowed to sleep. Otherwise, they would drown. PET scans of dolphin brains may help to determine how their architecture accomplishes this feat, another aspect of DARPA's PSD program. If those lessons are somehow applicable to human beings, we could see next-generation approaches to long-term wakefulness that rival modafinil, and if those methods are practical and their risks are limited, it will be hard to keep them out of the hands of civilians eager to get an edge in a competitive world.

Suppose that radically altered sleep patterns without threats to health were possible. What would be wrong with that? As Dinges notes, it's a debate we haven't had. The social implications of widespread use of modafinil's descendants would be enormous and difficult to predict. Libertarians would argue that government regulation would be overreaching, conservatives would worry about changing human nature, and liberals would worry about inequitable access to whatever advantages neuropharmacology might confer to those who are already relatively well off. All these views deserve to be aired.

LEANER AND COOLER

The military interest in altering normal biologic patterns to warfighting is hardly new. Longtime Minneapolis residents tell stories about the woozy, skinny young men seen about town during World War II. They were conscientious objectors in sleep and nutrition deprivation experiments. These days, DARPA is concerned about enhancing soldiers' capacity to go not only without sleep, but without food as well, and even to heal their own injuries. As reported by journalist Noah Shachtman, and described in DARPA's Web site, a project called "Metabolic Dominance" aims to develop a "nutraceutical," a pill with nutritional value that would vastly improve soldiers' endurance. (I'll confess that when I first saw the phrase "metabolic dominance," it conjured images of some especially odd sexual perversion.) The DARPA Defense Sciences Office explains the goal as "peak soldier performance":

> The vision for the Metabolic Dominance Program is to develop novel strategies that exploit and control the mechanisms of energy production, metabolism, and utilization during short periods of deployment requiring unprecedented levels of physical demand. The ultimate goal is to enable superior physical and physiological performance by controlling energy metabolism on demand. An example is continuous peak physical performance and cognitive function for 3 to 5 days, 24 hours per day, without the need for calories.

One idea is to get the body to switch on call from carbohydrate metabolism to lipolysis, basically relying on stored fats (ketosis) or, in other words, a highly efficient Atkins diet but, hopefully, without the risk that too much body fat will be used. That's not a problem for most of us, especially for short periods, but it could create risks for already lean young soldiers if they don't get lots of protein. Presumably, the DARPA program is intended to identify the most rapid results from the combined lipolysis switch and protein load.

When journalist Shachtman bounced the idea of substituting high-nutrition pills for food off some scientists, he got a mixed reaction. The chair of New York University's Nutrition, Food Studies, and Public Health Department wrote in an e-mail to Shachtman that "what this seems to be asking for is fantastic in every sense of the word. Calories are calories, laws of thermodynamics still operate, and humans are still human. I think they should use robots." (Fair enough; as we have seen, they might.) The Army has given out grants to see if herbs can enhance endurance and alertness. One candidate is echinacea, a plant that could be added to rations. Another approach is to adapt those nicotine-delivering transdermal antismoking patches to nutrient delivery. The Army's Natick Soldier Systems Center has also developed a high-energy meal consisting of three sandwiches, apple sauce with carbohydrates added ("zapple sauce"), and caffeinated gum. Yummy.

The agency has also invested in ways to see if the body's core temperature can be altered depending on weather conditions. Seriously injured soldiers might be able to go into hibernation while they healed, perhaps after self-administering advanced medication. Already, scientists have put a mouse into hibernation using tiny amounts of hydrogen sulphide, causing life processes to cease for six hours but then reversing the effect seemingly without injury. In principle, the technique could be extended to other mammals. "The ultimate goal," DARPA says, "is to enable superior physical and physiological performance by controlling energy metabolism on demand." A DARPA consultant told *Wired* magazine that "we're asking questions of biochemical processes that have been developed over eons. So there aren't going to be clinical trials tomorrow. But some aspect of this (regulating the body's internal heat) will be here faster than people think."

BUILDING SMARTER SOLDIERS

In his classic *Principles of Psychology* (1890), Williams James described the marvelous "plasticity" of the brain, referring to its ability to yield to new forces but not entirely. This quality of plasticity allows for the acquisition of new skills. The metaphor has proven to be of enduring value to neuroscience. Recent neuroscience also indicates that as a skill is being learned, the number of neurons applied to it and associated with it gradually decreases, adding a dimension of efficiency to the neural system.

It further turns out that at least one specific chemical in the brain probably inhibits plasticity, so that when it is decreased, learned behavior can be enhanced. NIH neuroscientists have found that when animals are working on new skills, their motor cortex secretes less GABA (gamma-aminobutyric acid), and that when they deliberately reduce GABA levels, the neural system is more plastic and the skill can be acquired faster. As Dartmouth's Michael Gazzaniga has pointed out in his book, *The Ethical Brain,* athletic and musical abilities could theoretically be enhanced by decreasing GABA, but no one knows what the long-term effects of such interventions would be.

Besides learning, another aspect of cognitive capacity that would be of immense value to combat personnel is an improved memory. Battle assignments can be complex and easily misremembered when fast-moving events unfold. Fighter pilots, for example, have to store a vast quantity of information in target-rich environments. One of the companies training monkeys to move a cursor only with their thoughts, Cyberkinetics, received DARPA funding reportedly because the agency is interested in increasing the bandwidth of soldiers' brains. A system called Braingate is being used with the monkeys but might eventually allow direct human connection to a computer memory and its reservoir of additional RAM for information-dense environments, like urban combat.

There are other silicon-based possibilities for memory enhancement. Dubbed a "brain prosthesis," a chip under development is intended to replace damaged parts of the brain. If it works for stroke or epilepsy, it might also be used to enhance normal brains. In memory-impaired people, the hippocampus, responsible for processing experience so it can be stored as memory in other sections of the brain, is often disrupted. An artificial hippocampus has been constructed in rats by electrically stimulating slices of the rat hippocampus and mapping which inputs yield which outputs. The resulting model can be encoded on a chip and placed on the area that needs help. Wires leading from one side of the compromised region to the device and from the device to the other side of the damaged area essentially bypass the compromised circuits. This simple-sounding idea took University of Southern California researchers ten years to accomplish, sponsored by DARPA, the Office of Naval Research, and the National Science Foundation. The USC team is testing its system on slices of rat hippocampus in vitro, and eventually will test it in monkeys to see if it changes behavior that involves memory tasks.

Biology offers other possibilities. Neuroscientists have found a gene that codes for N-methyl-D-aspartate (NMDA) receptors in the brain. When they gave adult mice extra copies of a type of NMDA receptor, the mice showed superior learning skills. Genomic and proteomic medicine may make this form of enhancement possible. But would individuals thus "enhanced" then be overloaded with memories, storing vast quantities of detail that are normally ignored because we have evolved to filter out or delete useless bits of information? In *The Mind of a Mnemonist,* the Russian psychologist A. R. Luria describes a patient who could not forget a single detail he had ever experienced. Unable to escape countless particulars, the man had to be trained how to forget; his absolutely retentive memory clogged his ability to pay attention to the present so that rather than appearing brilliant, he seemed timid and slow. Too much memory can literally be maddening, let alone counterevolutionary, unless the effects are short-lived. And who would want to volunteer for the first trial?

Thus, the introduction of a new memory storage system and bypassing our evolutionarily developed hippocampus raise the question whether our usual ability to slough off unneeded memories will be threatened, resulting in a cacophony of useless data that could drive one to distraction. Forgetting is often annoying but mostly adaptive, even a great relief. In the film *Eternal Sunshine of the Spotless Mind,* ex-lovers undergo a high-tech brain-erasing procedure to forget about the pain of their breakup. In a literally touching moment in *Star Trek,* Mr. Spock engages in an (unconsented) Vulcan mind-meld with Captain Kirk to help him forget a tragic love affair. Less romantically, undercover agents would benefit from the ability to lose their memories upon capture. Neuropsychologists have already found that deliberate memory loss among victims of parental abuse is both a demonstrable phenomenon (they are not "lying" when they say they don't recall) and a very effective defense mechanism. As the philosopher Bernard Williams has put it, "Forgetting is the most beneficial process we possess."

Experiments with monkeys may not give us the answer to these questions about the effects of a brain chip. The confusion associated with excess memory could result in subjective experiences that are not obvious to observers. There is as well a spectacular potential ethical problem in doing experiments with those whose hippocampus is damaged and can't form new memories: how could one obtain a valid informed consent from people who can't remember what they are doing, or why they are doing it?

Mucking around with memory also raises striking problems with personal identity. Going as far back as David Hume in the eighteenth century, philosophers have noted that our idea of ourselves is intimately bound up with our remembered experiences, including previous ideas about our selves that have entered the stream of consciousness. Anyone who believes that there are certain boundaries that should not be crossed must be concerned about the modification of abilities to remember and to forget. USC's brain chip that would divert electrical charges around damaged brain tissue presents the possibility for some personality change, for example, since a portion of neurons would be disengaged. But Oxford University philosopher Bernard Williams, who specializes in the theory of personal identity, has observed that this situation wouldn't be so different from excising brain tumors, which we have come to accept. Of course, even if the results were the same, the ethics of healing and enhancement could be different.

The artificial intelligence approach to enhancing and complementing natural memory would be more straightforward: engineer a direct connection between your brain and your Palm Pilot. Information could be not only uploaded to the brain but also downloaded to your Palm. DARPA's LifeLog program was a step away. As reported by Noah Shachtman, the idea was to create a database with every communication people have written, all pictures taken of them, and every bit of information about them, and then to use the Global Positioning System to track all their movements and sensors to record what they say, see, and hear, and add that information to the database. The unfolding events in a potential terrorist's life could be reconstructed in all their dimensionality. But so could yours or mine. The potential civil liberties issues presented by LifeLog got the attention of Congress, policy analysts, and human rights groups and was one of the reasons DARPA's budget was threatened with cuts in 2003. (LifeLog followed the embarrassing revelation of a planned program called Total Information Awareness, in which civilian records were to be used to identify potential terrorists; TIA didn't last long after it was disclosed.)

Citing changing priorities, DARPA quickly dropped the LifeLog project, disappointing artificial intelligence experts who were interested in developing the first models of how people collect and organize their experience. But in fall 2004, a more limited program called Advanced Soldier Sensor Information System and Technology (ASSIST) was created to collect everything a soldier experiences and does in combat. The public relations advantage of ASSIST over LifeLog is that it is limited to a soldier's experience, where privacy isn't a value that trumps information collection and analysis. But it does create a prototype that could have applications for a broader population and set of conditions. Information recorded in the ASSIST program might also be a concern for military lawyers responsible for defending soldiers in courts-martial: Will the data be anonymous? If not, what are the rules for its use?

SMARTER SOLDIERS THROUGH ELECTRICITY

Another approach to enhanced cognitive abilities for soldiers might lie in electrical stimulation of select brain centers. Evidence is accumulating that stimulation of some neurons as an adjunct to traditional rehabilitation can be of value for patients with paralysis. Nobody knows exactly why it works, but doctors at the Rehabilitation Institute of Chicago found that when they implanted electrodes in the motor cortex of stroke victims, they got significantly better results than with standard rehabilitation alone, recovering about 30 percent of lost function compared with 10 percent. Although the approach is not perfect, the gains for people whose arms had for years simply hung at their sides are wonderful. Some stroke patients with speech difficulties experienced improvement in that area, too, even though speaking ability was not the target of the experiment.

An intriguing question is whether electrical stimulation might help uninjured people exceed their normal intellectual capacities. After all, the brain operates on electrical energy. Could people acquire enhanced cognitive skills partly through neurostimulation? One technique being explored is called direct current stimulation or DC polarization. At the 2004 meetings of the Society for Neuroscience, NIH researchers reported that a tiny amount of electricity delivered to the brain through an electrode on the scalp (far less electricity than needed to run a digital watch) can produce measurable improvement in verbal skills. They ran the current through volunteers' scalps and asked them to name as many words as they could that began with a certain letter. The subjects showed about a 20 percent improvement when the current (two-thousandths of an ampere) was running. Since the current ran through the prefrontal cortex, the researchers speculated that the firing rate of neurons was increased, activating cells involved in word generation.

Though the volunteers' concerns about having their brains zapped were relieved after the scientists explained how tiny the charge would be, the associations with "shock therapy" will be hard for many to shake. (In fact, although electroconvulsive therapy [ECT] has acquired a bad reputation, it is often the only treatment that relieves both acute mania and acute depression, as in the case of one of my relatives who has suffered from a bipolar disorder for many years.) But DC polarization delivers a tiny fraction of the charge used in ECT and seems only to leave the subject with an itchy scalp. Of course, the fact that the technique does not involve surgery is also reassuring and makes it more practical than internally implanted electrodes.

Another noninvasive technique is transcranial magnetic stimulation (TMS), in which a magnetic coil is placed above the head and electrically produced magnetic pulses pass through the cortex. Depending on the particulars of the electrical signal that generates the magnetism, these pulses can alter the firing rate of certain neurons. As in DC, there is no pain from TMS, only the sensation of tapping on the skull as scalp muscles contract and a popping sound from the magnetic coil.

The therapeutic hope underlying these projects is of course that they can someday be used to treat stroke patients or those with dementias. TMS seems to be able to target specific brain regions more effectively than DC, but DC appears to carry less of a risk of inducing seizures. And, of course, the long-term effects of frequent exposure to electrical or magnetic stimulation are unknown. Nonetheless, DARPA has given grants to see if neurostimulation can improve impaired cognitive performance and reduce the other effects of sleep deprivation on soldiers, perhaps through helmets that deliver the tiny current.

Like so many potentially brain-enhancing technologies, neurostimulation can easily be oversold. Given how much we value cognition, however, even a modest improvement would be considered important. Long-term problems for military personnel might be hard to identify and could seem worth the risk for a marginal gain in mental agility in life-or-death situations. One Chicago scientist, Mark Huang, was quoted in the *Chicago Tribune* as observing that "there are many possibilities that have to be answered ethically. You can use [electrical brain stimulation] in any application where you want to potentially enhance brain function. If you want to learn a new language, potentially the stimulator might help. Would I recommend you do it for that purpose? No. But down the road, who knows? Obviously the sky's the limit and we're still in the infancy stage."

NO FEAR?

Speculative possibilities for "improving" soldiers by altering neural circuits and chemicals are endless and will no doubt be the subject of vigorous scientific, policy, and ethics debates in coming decades. Take another case: managing a gene for fear. A distinguished team of U.S. researchers reported in 2005 that a gene called stathmin, which is expressed in the amygdala, is associated with both innate (unlearned) and conditioned (learned) fear. The team bred mice without the gene (lab animals created without a certain gene are called "knockouts" for obvious reasons) and put them in aversive situations, such as giving them a mild shock at a certain point in their cage training. The normal mice exhibited usual fear behavior by freezing in place, but the knockout mice froze less often. That was the learned fear. When both normal and knockout mice were put in an open field environment, an innately threatening situation, the mice with stathmin spent more time in the center of the field and explored the environment more than the control mice.

Do people with lower levels of stathmin expression exhibit less fear? It's unlikely that there's any such one-to-one correspondence in humans, who are far more psychologically complex than mice and capable of modifying their genetically programmed behavior. Yet one can imagine that some imaginative military official who overestimates the importance of genetic information will someday propose screening Special Forces candidates, or even raw recruits, for the "fear gene." That someone would have this bright idea is not at all far-fetched. A few years ago, the Burlington Northern Santa Fe Railway Company had to pay $2.2 million to employees who had been secretly tested for a gene associated with carpal tunnel syndrome, even though the scientists who developed the technique himself said it couldn't work for that purpose. The company was trying to see if the workers' medical claims were due to their jobs or their genes.

If DNA testing for a fear gene is both scientifically and ethically dicey, what about setting out to create people who lack that characteristic? Would breeding humans without stathmin or any other genes associated with fear reactions in lab animals create more courageous fighters? That this conclusion would be a huge leap from animal studies might not stop some parents who harbor ambitions for a child capable of a glorious military career, or just don't want to give birth to a "sissy." Trouble is, fear or its functional equivalent is again one of those ancient properties exhibited by just about every animal. It surely has tremendous survival value so that its removal would be deeply counterevolutionary and would almost certainly generate numerous unintended and undesirable consequences for the individual, let alone thrust us headlong into a fierce debate about whether enhancing the human has gone too far. I'll have more to say about that looming question after we cover some more territory that illustrates some neuroscience-based options for enhancement, with military applications leading the way.

EASING EMOTION

The amygdala is an ancient organ that is critical to emotionality and memory. If the amygdala is damaged or removed, people may lose the ability to interpret cues from others that are intended to convey emotions such as anger. And it triggers release of the hormones epinephrine (also known as adrenaline) and norepinephrine into the bloodstream, which help emotionally weighted images to become firmly entrenched in long-term memory. All of us are quite familiar with the subjective experience, well described by the neuroscientist James McGaugh of the University of California, Irvine: "Whatever is being learned at the time of emotional arousal is learned much more strongly. . . . Any strong emotion will have that effect. It could be winning a Nobel Prize. It could be a very faint whisper in the ear, 'I love you,' at the right time."

If the amygdala's hormone-releasing processes could be inhibited, the searing of bad experiences into memory might be reduced. There is some evidence that beta-blockers, commonly used to treat heart disease, also have the ability to block neurotransmitters that consolidate emotion with long-term memory. Irvine researchers showed one group of volunteers a slide show that told a prosaic story about a boy and his mother. A second group was shown the same story except that the boy was hit by a car and his legs amputated. The members of a third group were shown the second, emotional story after taking a beta-blocker. When their memories were tested three weeks later, the drug group had flat emotional responses, similar to the group who had seen the uneventful version of the boy's outing with his mother.

A Harvard study of trauma victims had similar results. Some were given the beta-blocker propranolol; the others, a placebo. After a month of psychological counseling, scores on a post-traumatic stress disorder scale were lower for the beta-blocker group, but not significantly. However, after three months, 40 percent of the placebo group members had elevated physiological responses when they were asked to recall their traumatic experiences whereas none of the propranolol recipients did. With these experimental results as a clue, clinical trials of propranolol to prevent the chronic effects of traumatic experiences are already under way and show promise. Although the long-range burden of post-traumatic stress disorder for combat veterans is not strictly speaking a military concern, veterans and their families could be among the beneficiaries of effective drug therapy.

Neuropsychiatrists disagree about whether the promising results of early studies indicate that memories are being erased or that they are being bracketed so that they can be more easily handled by the sufferer. Whatever the actual mechanism, the therapeutic benefits of relieving painful memories are obvious in the case of people who have experienced trauma, such as combat veterans or the victims of sexual abuse. Those directly affected by terrorism, who psychiatrists say normally have a very poor psychological prognosis, could also enjoy relief. And who among us wouldn't prefer to be relieved of painful memories of terrifying accidents or the dull ache when we reach the anniversary of the deaths of those we have loved and prematurely lost?

But there are deep philosophical and perhaps sociological reasons that the use of these medications should at the very least be highly regulated. In its report *Beyond Therapy: Biotechnology and the Pursuit of Happiness,* the President's Council on Bioethics, which had studied the possible uses of propranolol in psychiatry and the implications of its wider use, raises some disturbing questions:

Would dulling our memory of terrible things make us too comfortable with the world, unmoved by suffering, wrongdoing, or cruelty? Does not the experience of hard truths—of the unchosen, the inexplicable, the tragic—remind us that we can never be fully at home in the world, especially if we are to take seriously the reality of human evil? Further, by blunting our experience and awareness of shameful, fearful, and hateful things, might we not also risk deadening our response to what is admirable, inspiring, and lovable? Can we become numb to life's sharpest sorrows without also becoming numb to its greatest joys?

The council's final judgment on the prospect of widespread use of drugs to blunt disturbing memory, a world with only happy memories, is worth pondering:

> To have only happy memories would be a blessing—and a curse. Nothing would trouble us, but we would probably be shallow people, never falling to the depths of despair because we have little interest in the heights of human happiness or in the complicated lives of those around us. In the end, to have only happy memories is not to be happy in a truly human way. It is simply to be free of misery—an understandable desire given the many troubles of life, but a low aspiration for those who seek a truly human happiness.

Soldiers and others who suffer from depression, insomnia, and other disorders following their traumatic battlefield experiences likely wouldn't find these philosophical reservations very convincing. Their concern, and that of their loved ones, is to reduce the torment of daily life. Nonetheless, it's worth considering the implications of these medications for the military before they are routinely included in field packs. Soldiers who could pop an antiguilt pill might not accrue experiences that lead them to hesitate when faced with an enemy they have been trained to annihilate. But military physicians have expressed appropriate concern about the force that might be produced by such a drug. My old friend and colleague Edmund Howe, who directs the medical ethics program at the Uniformed Services University of the Health Sciences in Bethesda, Maryland, and holds degrees in psychiatry and law, told New York's *Village Voice:* "If you have the pill, it certainly increases the temptation for the soldier to lower the standard for taking lethal action, if he thinks he'll be numbed to the personal risk of consequences. We don't want soldiers saying willy-nilly, 'Screw it. I can take my pill and even if doing this is not really warranted,

I'll be OK.' If soldiers are going to have that lower threshold, we might have to build in even stronger safeguards than we have right now against, say, blowing away human shields. We'll need a higher standard of proof [that an action is justified]."

LEARNING FROM OUR ANIMAL FRIENDS

Neuroscience isn't limited to human brains and nervous systems, of course. DARPA spokeswoman Jan Walker was quoted in *Mother Jones* magazine (no friend of DARPA to be sure) concerning the agency's interest in what can be learned from the sensory abilities of nonhuman animals: "We're interested in investigating biological organisms because they have evolved over many, many years to be particularly good at surviving in the environment . . . and we hope to learn from some of those strategies that Mother Nature has developed." In the same article, a DARPA project manager added that "inspiration from nature . . . will allow more life-like qualities in the system."

A 2001 report by the National Academy of Sciences (NAS) on new opportunities in biotechnology and information technology described several DARPA initiatives that reflect this principle of learning from nature. One is called the "electronic dog's nose," attempting to understand how dogs' neurosensory processes are able to detect explosives and use that information to develop a model for electronic bomb sniffers. Besides modeling artificial systems based on animals, there's also the possibility of altering the animals themselves. A type of wasp larvae can be exposed to certain vapors so that when it matures it detects explosives or "odors of interest." Another idea is to install electronic chips in insects "so that their hunting patterns become search algorithms for DoD sensors," in the words of the NAS report. Evolution has taught insects the most efficient ways to scan their environment, techniques that could be useful in the design of human sensing systems.

Other nonneuroscience animal-inspired DARPA projects include studying how geckos climb walls, how octopuses hide, and how various critters employ adaptive camouflage. The *Mother Jones* article noted an apparent "animal fetish" as even projects that don't have a particular link to our lesser friends have been given monikers such as "Big Dog" (a robot canine), "WolfPack" (a group of miniature ground sensors), "Piranha" (enabling submarines to engage elusive targets), and "Hummingbird Warrior" (producing a vertical takeoff and landing unmanned vehicle). In fairness, the auto industry also likes animal names (mustang, skylark, impala, etc.). They have a way of rendering imposing machines more approachable, even personal.

The study of the perceptual systems of other animals not only will stimulate new designs for sensing devices to aid human operators, but will certainly lead some to wonder about the feasibility of introducing genes or proteins to modify human beings. We already introduce biological materials of other animals into the human body (vaccines, porcine valves for hearts), and are already going down the road of genetic modification of prospective children (prenatal testing and screening, genetic engineering), so why not consider putting genes in adults as well? As is generally the case when national security is at stake, risks to the recipients are likely to be assessed differently than they would be in the normal context of medical care or research. But should they be? Thus, we are once again led to ethical issues in enhancement technologies.

THE ETHICS OF ENHANCEMENT

Should we build better soldiers through "artificial" enhancements? Is there even a valid distinction to be drawn between artificial and "natural" enhancements such as exercise and discipline? Aren't we just trying to gain whatever advantages we can as nations have always tried to do, or are these techniques cheating nature? Can we manage the consequences, or are the risks for the individual and for our society too great? These questions are part of a raging debate about whether we should use new discoveries in neuroscience and other fields like genetics to improve ourselves, our descendants, and perhaps even the species. If it would be acceptable to enhance civilians, then it's hard to see why national security agencies should be barred from giving warfighters an edge. If it's not acceptable to enhance civilians, there still might be a special case to be made for tuning up soldiers, but the argument for a military exception will need to be a powerful one. So the more general enhancement debate is important for the idea of building better soldiers.

There are special features of the enhancement debate in the military context. Under normal circumstances, individuals can of course refuse to do things that other people think will "improve" them. Workers and students can decline to accept enhancements that their employers or schools recommend, perhaps at the cost of losing their positions. But military personnel might not have that privilege. According to the Uniform Code of Military Justice, soldiers are required to accept medical interventions that make them fit for duty. Experimental treatments are a harder case, but the U.S. government has shown a tendency to defer to commanders in a combat situation if they think some treatment is likely to do more harm than good, even if unproven. An example is the use of the anthrax vaccine during the 1991 Gulf War, though it had not been shown effective for inhalational anthrax in humans. However, as Evan DeRenzo and Richard Szafranski observed in an Air Force magazine article on the ethics of human performance enhancement in the military, freedom-of-choice arguments have more traction in an all-volunteer army than in a force of draftees.

I trace the modern debate about the general ethics of enhancement to an esoteric discussion that took place in the pages of philosophy of medicine journals in the 1970s. The issue was the meaning of health and disease, two concepts that are very familiar but surprisingly hard to pin down. What exactly does it mean to be healthy or sick? There seemed to be general agreement that the concepts of health and disease turned on the idea of normalcy. Some argued for a statistical sense of normalcy, others for the notion that to be normal is to be capable of "species typical functioning" with regard to survival and reproduction. No refinements seemed able to avoid the fact that the ideas of health and disease (literally "dis-ease") are useful intuitive guideposts but not very precise; nor are the concepts of treatment and enhancement. Take the example of advanced sleep medication for males in their forties, a time of life when many men experience disrupted sleep patterns. Assume that some degree of insomnia is typical for human males beginning in their fourth or fifth decade. Is Ambien a *treatment* for a sleep disorder or an *enhancement* to sustain a more youthful sleep schedule?

The enhancement debate picked up steam in the 1990s with the spread of drugs like Ritalin and Prozac; then, the steroid scandal in sports and the erectile dysfunction treatment craze made the issue nearly inescapable. On one extreme are the transhumanists, a philosophical movement populated by scientists, philosophers, futurists, and others that has grown in recent years with national meetings and publications. The leading transhumanist writers call explicitly upon neuroscience to depreciate the boundaries between treating disease and enhancing the individual. In *More Than Human,* Ramez Naam cites work on computer chip neural implants as an example of the emerging capacity to go well beyond healing the sick to vastly improving what are thought of as normal capacities. Computer hardware outside the body's boundaries will be unnecessary, according to Naam, as implants will enable us to Web surf, send e-mails, and enjoy far more powerful sensations simply by intending to do so. In *The Transhumanist FAQ,* Nick Bostrom similarly contends that human evolution is still primitive, "that the human species in its current form does not represent the end of our development but rather a comparatively early phase." We as a species can alter ourselves by the technology we have created, and that very technology "will eventually enable us to move beyond what some would think of as 'human.'" That posthuman or more fully human creature will be perpetually young and healthy and vastly more capable of experiencing love, beauty, and tranquility than are we. Ultimately, we will be able to upload our very selves to new bodies as needed or desired, having downloaded all the contents of our mind to hugely powerful computers with multiple backups to ensure the self's immortality.

On the other side of the enhancement argument are many neoconservative thinkers like Francis Fukuyama, a Johns Hopkins University political philosopher who also happens to be a member of the President's Council on Bioethics. Generally considered to be dominated by neoconservatives, the council published the report I quoted earlier called *Beyond Therapy* that expressed reservations about memory control. The report also expressed skepticism about the enthusiasm of many Americans for any drug or device they can get that seems to offer easy self-improvement. Fukuyama is about as upset as he can be about the unbridled enhancement philosophy of the transhumanists. In a 2004 article in *Foreign Policy* magazine, Fukuyama called transhumanism the world's most dangerous idea. Transhumanism seems reasonable "when considered in small increments," Fukuyama writes, which is exactly what makes it so dangerous. "It is very possible that we will nibble at biotechnology's tempting offerings without realizing that they come at a frightful moral cost."

In his book *Our Posthuman Future,* Fukuyama expresses two main worries about the transhumanist goal. First, if we ever succeed in creating beings with far greater abilities than ourselves—through biotechnology or genetics or neuroscience or whatever—political equality will be jeopardized. By tinkering with our essential human nature, the universal essence of humanity will have been changed, and with it the rational basis for thinking of all persons as equal in the political system. Second, the intricate human being created through millions of years of evolutionary selection is carefully balanced between, for example, violence for self-defense and affection for social cohesion, and deliberate interventions are unlikely to achieve creatures with just the right blend of good and bad qualities. Nor can we necessarily discern what those qualities are, as they have to prove themselves in the complexities of social life with its countless variables.

I find myself squarely in the middle. I'm not as sanguine about a hyperenhanced future as the transhumanists, nor am I comfortable with their utopianism. As David Hume wisely observed, the future tends to resemble the past. We can expect a range of consequences as we incorporate new technologies and instrumentalities into our lives and bodies. I also have doubts about the metaphysics of the idea that the subjective experience of personal identity can ever be captured, even by the most comprehensive memory chip. But neoconservatives such as Fukuyama seem to me to harbor an excessively dour view of the technological future. First, I'm not as convinced as he is that the idea of human equality is grounded in a universal concept of essential human nature; my more pragmatic view has it that human equality is a rather squishy moral notion. It feels right to most of us, and we rally around when we need to. Second, there is plenty of room for argument about his view that human evolution has produced a mix of good and bad qualities that can't be improved upon. Still more fundamentally, I can't swallow the suggestion that, in a world of ethnic and religious tension, nuclear proliferation, global warming, emerging infections, and terrorism, transhumanism is our biggest problem.

Where I do emphatically agree with Fukuyama is that the proper response to transhumanism is not to prohibit research and development of these new technologies but to develop careful monitoring and regulatory systems. Some of this can be accomplished by the scientific community and its organizations. For example, in 2005 a committee that I cochaired recommended guidelines on human embryonic stem cell research. The committee was created by the National Academies, an organization of elected members that is chartered by the federal government to advise it on science, medicine, and engineering issues. Because the Bush administration has allowed only limited federal funding for research involving human embryonic stem cells, scientists weren't sure what research would be considered appropriate, especially since several states and private companies intended to do work involving this controversial field. Among many other recommendations, our committee concluded that no human embryonic stem cells should be placed into nonhuman primates at any stage of development. Part of the concern is that some of the human cells might turn into brain cells in, say, a rhesus monkey embryo, and they might contribute in an organized way to the monkey's brain. While we can't know if the monkey's brain would be changed by the human cell contribution, we also can't be sure it wouldn't be. Would any resulting creature feel like a human "locked" in a monkey's body? While highly unlikely, this possibility arguably carries a serious ethical burden. For similar reasons, we also recommended against putting embryonic stem cells from other animals into human embryos.

Given the publicity and sensitivity of the embryonic stem cell issue and the prestige of the National Academies, we felt sure our recommendations would be adopted by legitimate research centers and individual scientists. But that voluntary arrangement falls well short of government regulation. Also, with regulation there often comes significant government funding for research, which acts as an important incentive to follow the rules. We have seen how important government funding is in keeping a new area of medical science on track in the case of in vitro fertilization. When the Reagan administration decided to stop funding for IVF research in 1980, the emerging industry was cut loose without public scrutiny. The result was what many consider to be a field that bears a resemblance to the Wild West, with all sorts of practices pursued and claims made and with only limited public scrutiny and modest (and relatively recent) self-regulation.

National security research on enhancement technologies will require the close involvement of advisory bodies of people both in government and outside it, with as much transparency as possible and, when transparency must be limited, with whatever security clearances are needed to make full participation possible. While some general principles should be articulated and become part of our regulatory framework, much of the hard work will have to be done on a case-by-case basis. There are some models out there for ethical review in security policy that I will talk about in the last chapter. The ethics of enhancing warfighters' capabilities with emerging neurotechnologies needs to be moved onto our national policy agenda.

7 :

E N T E R   T H E   N O N L E T H A L S

IN 'THE ILLIAD,' HOMER CELEBRATES THE RAW COURAGE exhibited in
one battle: "They did not fight at a distance with bows and javelins, but
with one mind hacked at one another in close combat with their mighty
swords and spears pointed at both ends." Whatever its manliness, over the
centuries the need for hand-to-hand combat has gradually been eroded.
From catapults to cannon to manned bombers to unmanned drones to
satellites, tactics have shifted from literally "standing up" to an enemy to
evasion and destruction. Commanders who resisted these technical advances have been left behind by history. Legend has it that when General
George S. Patton told journalists after World War II he regretted that the
emergence of airpower would deprive soldiers of the opportunity for her-
oism, he got in hot water with his superiors. Accurate or not, the story sig-
nifies the way warfare has changed.

A MATTER OF HONOR?

When we think of more distant forms of weaponry, we tend to think
in terms of the increasing violence and overwhelming force of measures
such as aerial bombardment and the atomic bomb itself. We also associate
these "advances" with the greater likelihood that "collateral damage" will
occur and with the growing tendency toward total wars in which civil-

ian casualties not only are coincidental but actually can be viewed as part of the tactical effort to demoralize the enemy, such as the firebombing of Dresden and the bombing of London.

But along with these "harder" forms of warfare have also come "softer" or less-than-lethal forms. The varied attempts to take advantage of human psychology in warfare that have accelerated in the last hundred years, including psychological operations and propaganda, could be regarded as weapons of war not intended to kill or injure. Traditionally, the most desirable result of these forms of intimidation was capitulation by the cowed adversary, however unlikely. If in spite of these efforts the war was reduced to the most primitive level of direct physical contact, the psychological softening up might at least have done just enough damage to enemy morale to create an advantage. The extended siege was another historic nonlethal option, though massacres of those within walled cities sometimes followed. Various neuroscience-based innovations will considerably lengthen the menu of less-than-lethal warfighting options.

IN THE THEATER

Drugs that affect the brain and that are widely used in medicine are being evaluated for other purposes. Many of the agents I will describe in this chapter can in theory be applied to police situations such as hostage taking in which it is desirable to avoid death and permanent injury. But a particular incident educated many people to the problems of using drugs in uncontrolled settings. On October 23, 2002, over seven hundred people attended the musical *Nord-Ost* in a Moscow theater with the clunky Soviet-era name House of Culture of the State Ball-Bearing Plant Number 1. During the second act, forty terrorists demanding Russian withdrawal from Chechnya stormed the building, taking patrons and actors hostage. The terrorists made it clear in a videotape that they would die rather than surrender. Conditions over the next several days were miserable as the hostages, including dozens of children, were deprived of food and water. The terrorists spread explosives throughout the building to ensure catastrophe if they were attacked.

Early on October 26, Russian commandos piped an anesthetic gas called fentanyl through a hole in the wall, hoping to incapacitate the hostage takers. The action succeeded in putting many of the terrorists to sleep and disrupting the rest. All of the Chechens were killed, some shot at point-blank range. Although dozens of ambulances were standing by and hospital emergency rooms were at the ready, the medical personnel were unprepared for the problems they actually faced—not injuries due to bullets or shrapnel, but unconscious men, women, and children. In a horrible lapse of planning, authorities had not told the emergency workers what type of gas had been used, though fentanyl is a familiar agent in operating rooms and its effects could have been managed. As it was, 128 people died of the combined effects of the fentanyl and dehydration, with disproportionate casualties among the children, whose small bodies were more easily overwhelmed by the drug.

The incident and its consequences led to severe repercussions. In response, President Vladimir Putin declared a national day of mourning and decided to step up Russia's grip on the Muslim province. In 2004, hundreds more Russians were killed in multiple suicide attacks, including two downed passenger aircraft, as part of the continuing Chechen terrorist resistance.

In an unfortunate bit of timing, in Washington on the same day as the unfolding tragedy in Moscow, the National Academy of Sciences released a report on the prospects for effective military use of "non-lethal weapons," including "calmative" agents like fentanyl. The academy's committee concluded that the Chemical Weapons Convention was ambiguous enough to permit the use of some nonlethal chemical weapons. Among the recommendations of the report, entitled *An Assessment of Non-Lethal Weapons Science and Technology,* was that the Pentagon's Joint Non-Lethal Weapons Directorate (more about the directorate later) should establish "centers of excellence" in the weapons' development. In 2004, the Pentagon's Defense Science Board, in a report entitled *Future Strategic Strike Forces,* observed in a similar vein that "calmatives might be considered to deal with otherwise difficult situations in which neutralizing individuals could enable ultimate mission success." Likely targets: "when striking rogue or terrorist leadership, the mission is to kill the leaders themselves" and "to decapitate regimes." The Defense Science Board is only an advisory committee and doesn't set Pentagon policy, but having been on a few government advisory committees, I can certify that these groups don't generally issue lengthy reports for their own entertainment. The board's comments likely give a clue to the direction of Pentagon thinking.

Proponents of "nonlethal" weapons (NLWs) claim that they will obviate the need to kill or maim. These weapons are actively being sought by all branches of the U.S. military and come in a dazzling variety of forms: calmatives or "incapacitants"—chemicals that put people to sleep; acoustic and light-pulsing devices that disrupt cognitive and neural processes; odors so disgusting they sicken; sudden colored fog that creates panic; optical equipment that causes temporary blindness; and mechanisms that stimulate nerve endings as though they are on fire, among dozens of others. A striking fact about this list is that all are related to the human brain and nervous system. In the rest of this chapter, I'll describe and discuss the implications of some of these NLWs, after some more orientation to their history and the rules surrounding their development and use.

TALKING NONLETHAL

Since the mid-1990s, the U.S. Marine Corps has been the home of the Joint Non-Lethal Weapons Program (JNLWP). The JNLWP defines nonlethals as "weapons that are explicitly designed and primarily employed so as to incapacitate personnel or materiel, while minimizing fatalities, permanent injury to personnel, and undesired damage to property and the environment." According to the program's official history, contemporary interest in a systematic approach to nonlethal weapons development started when the top brass was interested in using NLWs in the withdrawal of United Nations forces from Somalia in 1995. Although they were never used in Somalia, the idea that such weapons could be useful in "Operations Other Than War" ("OOTW" in Pentagon parlance), such as in situations of urban insurgency, was given a boost by this episode. According to GlobalSecurity.org, a Web site devoted to news and security information, "Non-Lethal munitions applications will be used by military personnel to apply the minimum force necessary while performing missions of crowd control and area security at key facilities around the globe. These devices will aid military forces/commanders in situations of hostages rescue, capture of criminals, terrorists, or control of other adversarial persons."

Discussing nonlethal weapons presents two general problems. The first is the name. Some governments and agencies prefer terms such as "less than lethal" or "less lethal" to "nonlethal," since in sufficient quantities or delivered a certain way or to the wrong people (the children in the Moscow theater), just about everything can be lethal, including iced tea. The second problem is the vast array of technologies that can fall into this category. One Canadian analyst has suggested the following typology, which I directly quote:

a. Antipersonnel
   1. Physical; rubber/plastic and beanbag rounds, foam batons, nets, water cannon, etc.
   2. Chemical; CS gas, pepper spray, sticky foams, olfactory agents, calmatives, etc.
   3. Directed energy; flash-bang grenade, stun gun, eye-safe laser, loud audible, etc.
   4. Biological; no legal antipersonnel agents

b. Antimateriel
   1. Physical; vehicle nets, fiber and wire entanglements, caltrops, etc.
   2. Chemical; sticky foams, combustion modifiers, metal fibres, friction reducers, filter cloggers, super-corrosives, super adhesives, etc.
   3. Directed energy; pulsed power, high-power microwave, particle beams, infrasound, ultrasound, etc.
   4. Biological; biodegrading agents for petroleum products, rubber, explosives, etc.

Obviously, only some of these items are designed to act directly on the brain and nervous system, and fewer still build on theoretical breakthroughs in neuroscience. But some do, and they can be considered to fall within the ambit of brain science and national security. Among the antipersonnel weapons in this typology, those that involve the nervous system include certain chemicals, especially calmatives such as fentanyl, and some directed energy weapons, such as beamed microwaves.

Growing concerns about terrorism have fed interest in NLWs. Contemporary arms stockpiles have typically been designed for fighting between nation-states. The use of conventional nonnuclear and nuclear weapons in the places terrorists like to operate would result in high levels of noncombatant casualties that may be politically as well as morally unacceptable. These kinds of situations are often referred to as "asymmetric" conflicts in which the force used in response to, say, a car bombing should not be so great as to stimulate sympathy for insurgents. Many defense experts have observed that in the new global political environment, various levels of force need to be available. Hence, defense planners' interest in NLWs. GlobalSecurity.org summarizes the rationale for NLWs:

> The temporary discomfort and confusion generated by some of these Non-Lethal munitions provides the tactical team the few seconds necessary to exploit the situation by redirecting the actions of a targeted individual or group and enhances the ability to apprehend same. The shade of light green has been selected to be the ammunition color-coding for all Non-lethal ammunition components. Non-Lethal devices are intended to confuse, disorient, or momentarily distract potential threat persons. They are designed to produce only temporary incapacitation to either innocent bystander or threatening individuals.
>
> Minor injuries can and will occur (bruises, stings, etc.) to individuals who are struck by payloads of Non-Lethal munitions. In fact, even if properly employed severe injury or death are still a possibility. Non-Lethal weapons shall not be required to have a zero probability of producing fatalities or permanent injuries. However while complete avoidance of these effects is not guaranteed or expected, when properly employed, Non-Lethal weapons should significantly reduce them as compared with physically destroying the same target.

CALM DOWN

When I read about a contract for a nonlethal mortar weapon lent by the Marine Corps Research University to Penn State researchers, I called a friend in military research to ask him what a nonlethal mortar would do. "I think," he said mildly, "it would be something that would put you to sleep," precisely the intent of the Russian troops outside the House of Culture. If I'd been following the Soviets' Afghan war more closely, I might have noticed reports that calmative agents were said to have been used there against the Mujahideen. Though these reports have never been confirmed, they do become more interesting in light of the decision to use an anesthetic gas in the theater. Presumably some people in the ex–Soviet Russian military had the expertise, or thought they did.

Calmatives include a large class of psychoactive drugs that can cause hallucinations as well as drowsiness, and compounds that depress or inhibit the function of the central nervous system. They include drugs such as alfentanil, fentanyl (used at the Moscow theater), ketamine, and BZ. Some of them can be mixed with dimethyl sulfoxide (DMSO), increasing the normal absorption rate through the skin into the bloodstream by about 1,000 percent.

Parallel to the growth in the underlying neuroscience, calmative research in the United States appears to have accelerated in recent years. A 2000 report by a team at Penn State's Applied Research Laboratory observed that "since the mid-1960s, the availability of these pharmaceutical agents . . . have [sic] undergone a remarkably rapid phase of growth. Indeed, the premier status of the US pharmaceutical Industry [in] the world markets, combined with the exponential developments in the fields of pharmacology, neuroscience, anesthesia, and biotechnology fields, among others, has brought forth a diverse array of compounds that produce sedation and/or a calm state as either a primary or secondary effect." Among the major classes of drugs listed in the report are sedatives, anesthetic agents, muscle relaxants, opioid analgesics (chemical relatives of morphine that are taken for acute pain), antianxiety agents, antipsychotics, and antidepressants. "In seeking to identify pharmaceutical agents useful as calmatives in a non-lethal technique, several characteristics may contribute to the profile of an 'ideal' agent. The calmative should be easy to administer and adaptable for administration via topical, subcutaneous, intramuscular, or oral route. The onset of action for this compound should be fast (seconds to minutes) and most likely of short or of a limited duration (minutes)."

The Sunshine Project is an organization based in the United States and Germany that tracks developments in bioweapons and works "against the hostile use of biotechnology in the post–Cold War era," according to its Web site. Though it certainly has an antiweapons and left-wing bias, I have found the group's work reliable, including a report that the Joint

Non-Lethal Weapons Program engaged in a war game "to identify alternate means of offensive operations that will provide the National Command Authority (NCA) and Joint Force Commanders (JFC) additional operational options when executing a coercive campaign." The U.S. Army has funded a project at a New York pharmaceutical firm to aerosolize an anesthetic called ketamine, and a similar project conducted by the same company was to take place at Johns Hopkins University. These efforts are yet another example of the dual use phenomenon, because they are aimed in the first instance at medical uses of ketamine but will surely provide data for its use as a nonlethal weapon and perhaps information about similar applications of other anesthetic agents.

THE SOUND OF SILENCE

Acoustic devices present many nonlethal possibilities. At a primitive level, in 1990 American forces played loud music to drive the deposed dictator Manuel Noriega so stir-crazy that he walked into the arms of waiting American soldiers outside the Vatican embassy in Panama where he had been granted diplomatic cover. One wonders how eager the local Vatican officials were to see him go at that point, since he was the only one who could get the music to stop.

More subtle acoustic devices are now being developed and even field-tested, some of these raising questions about mind control and manipulation. When I was growing up, there was a lot of interest in—and considerable controversy about—subliminal video signals allegedly being conveyed in TV commercials. Supposedly these were milliseconds-long snippets of visible text encoded in advertising that were indiscernible to the conscious mind but worked on the unconscious level to influence the viewer's buying habits. If Madison Avenue could do this, then so could the Communists, eating away at our freedom while we watched a soap ad for *The Beverly Hillbillies.*

The popular legend about subliminal advertising had its roots in the more general 1950s preoccupation with brainwashing. But suppose you were the Tom Cruise character in the film *Minority Report,* set in 2054. Whether the screenwriters knew it or not, that's exactly one hundred years after the beginning of the CIA's MKULTRA program that tested hallucinogens on unwitting subjects. Strolling through a shopping mall, the Cruise character, Officer John Anderton, hears advertisements designed to entice him to make various purchases. The interesting twist is that the inviting commercials ("Hi John, welcome to The Sharper Image. . . .") use his name (probably picking up some identity chip on a credit card he was carrying), and only he can hear them.

Flash back to Tokyo 2005, when pedestrians passing Coke machines on the street hear the familiar "pssst" of a carbonated beverage can being opened and the clinking of ice cubes in an empty glass. Again, only that passerby hears the ad, carried on a beam of sound so narrow that someone a foot away hears nothing. This is hypersonic sound (HSS), developed by a brilliant, autodidact inventor named Woody Norris, whose American Technology Corporation (ATC) is refining the revolutionary audio technology into dozens of uses, civilian and military. Revolutionary is not too strong a word: the basic design of conventional speakers hasn't changed in more than eighty years, but *USA Today* has compared the promise of hypersonic sound to the contrast between a jet and a propeller-driven aircraft. The comparison is apt, though it might well be conservative.

The idea of hypersonic sound is one that others had been working on for a long time but that Norris accomplished. Normal "linear" sounds create pressure waves in the air. Change the frequency of the wave to one that is ultrasonic and it can't be heard by the human ear. Narrow the path of that ultrasonic wave and mix that energy with the air in that path, and you get a column of sound that does not spread out like conventional sound but stays locked like a sonic laser in that pathway. Put your ear in that column and you'll hear the sound, or bounce the column off a hard surface and you'll hear it on the rebound. Norris' company designed a signal processor, an amplifier, and a transducer that emits the sonic energy, actually two ultrasonic information-laden waves that bump into each other when they encounter a solid object. That causes the air around the listener to change and re-create the original sound. Some describe the sound as seeming to be right at the ears, or even a bit inside, like the most high-fidelity headphones.

Hypersonic sound might be applied anywhere that it would be advantageous to direct a message to a single individual in a group. The kids can listen to their music in the backseat while Mom and Dad listen to theirs in the front, each undisturbed by the other. Patients sharing a semiprivate hospital room could enjoy separate audio from their television sets. A museumgoer can hear the display narration by standing in the right spot, while the next person hears the previous or successive narration. Out of tens of thousands of onlookers in a football stadium, a message— "Hi. Nice orange and green striped shirt, sir. Ready for your next hot dog? Don't forget the Coke!"—could be beamed to one individual. At home, surround sound effects can be accomplished by bouncing the signal off a rear wall with hypersonic speakers on the TV, without wiring or even rear speakers, and of course without bothering anyone else in the house. Here's a partial list of other applications from the ATC Web site:

- Automobiles—HSS announcement device in the dash to "beam" alert signals directly to the driver
- Audio/Video Conferencing—project the audio from a conference in four different languages, from a single central device, without the need for headphones
- Paging Systems—direct the announcement to the specific area of interest
- Retail Sales—provide targeted advertising directly at the point of purchase
- Drive Through Ordering—intelligible communications directly with an automobile driver without bothering the surrounding neighbors
- Safety Officials—portable "bull horn" type device for communicating with a specific person in a crowd of people

By coincidence, not long after I initially learned about hypersonic sound, a journalist called to ask if I thought the technology was entirely benign (she didn't ask about the military applications, which I'll get to soon). I responded that as with most technologies, I found some of the applications quite marvelous, some annoying, and some worrisome. I am delighted at the prospect of enjoying my favorite action movies in surround sound full blast without my wife complaining; she's an excellent cook who likes to try the latest recipe in silence. On the other hand, as a dedicated shopping mall crawler, I will regret the seeming loss of anonymity while "just looking," not to mention the sheer annoyance of having cloying advertising voices invading my head.

But as one who has known many people with mental disorders, what concerns me far more about widespread use of hypersonic sound is the risks to those who have difficulty distinguishing fantasy from reality. As I told the journalist who called me about HSS, our basic concept of reality is closely tied to intersubjective experience. If we hear an odd noise we will, almost without thinking, turn to a companion and say, "Did you hear that?" If hypersonic sound comes into common usage in commercial and other contexts, those deep intuitions will be challenged by everyday life.

For most of us this won't pose a problem, but for a highly vulnerable few, including those with some forms of schizophrenia, confusion caused by hypersonics could be life threatening. Afflicted persons who are often able to keep their grasp on reality and distinguish hypersonic from non-hypersonic sounds may have their paranoid tendencies stoked by the fact that such silent messages are in fact so easy to deliver. Conspiracy theorists will have a field day, as some of their worst fears will have been actualized not by the CIA, but by incessant commercial promoters. As Wrye Sententia, the director of the Center for Cognitive Liberty and Ethics, pointed out to me, hypersonic sound may pose greater problems than brain mapping. "HSS involves introducing something into cognition that you can't close yourself from. You can close your eyes but not your ears. The issue of how our senses interact with our thinking is going to become an increasingly significant regulatory concern."

My guess is that these very problems with hypersonics may cause strict legal limits on its use in public spaces. Even if legal constraints are not imposed, the sheer annoyance of the individualized invasion of mental space may prove to be a disincentive to advertisers who prefer not to alienate their potential customers. Hypersonic sound around every corner would be noise pollution run amok. Of course, it is possible that we will all become inured to this invasion, as we have to the daily bombardment of uninvited visual messaging. But sound is a more intimate sense than sight (deafness is said to be a more isolating disability than blindness), so even imaginative capitalists might have trouble overcoming the psychology of hearing. Admittedly, though, hypersonic devices have many constructive potential uses, such as targeted messages to the blind at crosswalks. Regulation rather than prohibition seems to be the destiny of this technology.

Whatever the outcome of the coming debate about hypersonic sound in the civilian context, the technology seems to have a bright future in the realm of military applications. The U.S. Army has spent millions of dollars on long range acoustic devices (LRADs), already scheduled for installation in the new Stryker fighting vehicle. The LRAD uses a hypersonic beam to hail or deliver warnings to individuals more than three hundred yards away. The Third Infantry Division in Iraq is said to be planning to use LRADs in its security operations, maintaining order in cities where insurgency is a significant challenge. As an American Technology Corporation press release notes, "Recorded messages can be selected and transmitted over LRAD in multiple languages. Used in land-based roles for military operations other than war, LRAD will support missions including crowd control, area denial of personnel in checkpoint operations, and clearing buildings." For use on noisy and dangerous aircraft carrier flight decks where communication is a familiar problem, the Navy will adopt American Technology Corporation–designed speakers, and LRADs will be installed on various vessels, military and commercial, for long-distance hailing and warnings. An Aegis Destroyer under construction at Bath Iron Works will be among the first Navy ships to test this technology.

A more aggressive form of hypersonic sound with other implications for security is a system known as high intensity directed acoustics (HIDA). HIDA produces a "sonic bullet" that can create pain so intense that it can be disabling, causing the recipient to lose balance, to vomit, and to develop a migraine. A recording of a baby's cry played backward is a very effective vehicle for these purposes, it turns out, even when not played at the highest HIDA volume. (As countless generations of new parents have noticed, evolution has been quite clever in designing this way of infantile attention-getting.) And again, it's important to remember that HIDA's utility as a form of HSS is that only the target experiences it. Beamed at terrorists or used as an interrogation tool, HIDA seems to qualify as one of a large number of NLWs as it disables but does not kill; it may not even leave evident long-term injuries.

HIDA is not entirely new. The British army tested a "squawk box" in Ireland in 1973 that mixed two ultrasonic frequencies and caused nausea or fainting when they reached a human target. The Russians are thought to have developed a high-power, low-frequency device that emits an acoustic bullet from small antenna dishes. The waves in front of a target can cause injury and even death. It has long been known that high-intensity strobe lights can have a similar effect working through the optical rather than the auditory system, causing dizziness, disorientation, and nausea. They can also be used for crowd control, but in some people they may trigger epileptic seizures.

THE SMELL TEST

One especially bizarre brain-related category of NLWs is malodorants, otherwise known as stink bombs. First tried during World War II, stink bombs seem to have captured the imaginations of kids growing up in the 1950s and 1960s; I remember hearing the term used by my peers. Fecal odors have been a long-standing preoccupation of malodorant weapons designers. During WWII, the idea was to covertly rub a chemical mixture called "Who Me" on German occupiers to make them "the object of derision," according to a 1998 military document.

The notion that different ethnic and racial groups react to odors differently was of special interest during the Vietnam War when it would have been attractive to flush guerrillas out of their tunnels, bunkers, and other hiding places with a weapon that would not nauseate American troops. In 1966, a DARPA-funded study at Ohio's Battelle Institute tried "to determine whether intercultural differences in olfaction exist, particularly with respect to offensive smells, and if they do, to what extent they can be utilized in psychological warfare." Anthropological writings on indigenous Asian peoples were examined for clues. Another idea was to use classical behavioral conditioning to associate antipersonnel bombs with a certain odor so that eventually only the smell would be needed to sow panic.

Based on information contained in publicly available documents, ethnically targeted malodorants appeared to be the subject of renewed interest at the U.S. Army Edgewood Chemical Biological Center in Maryland in the late 1990s. A 1998 service order obtained by the Sunshine Project under the Freedom of Information Act claims that standard bathroom odor is an ineffective weapon because "it was found that people in many areas of the world do not find 'fecal odor' to be offensive, since they smell it on a regular basis." Therefore, the document continues, "the objective of this work is the development of a comprehensive set of [malodorants] that can be applied against any population set around the world to influence their behavior." To enhance crowd control capability in different cultures, the contract seeks results of exposure to various malodorants "based on a diversity of geographic origins and cultural heritage." The resultant "odor response profiles" could be applied to "elicit a favorable behavioral response" among different groups, in other words, incapacitating them, causing them to panic and flee. A draft report from 2000 states that "U.S. Standard Government Malodor" was tested on culturally diverse volunteers from the United States and South Africa. Sewage, vomit, and burnt hair odors were among those tested, compared with several pleasant smells such as cherry and cinnamon to ensure the participants had normal olfactory sense.

The Sunshine Project also reports that efforts to generate the most offensive smell possible have attained new scientific heights. In June 2001, a Texas company headed by a former Navy commander and Naval Laboratories scientist took out a patent on the compound that makes feces smell, noting that "the use of obnoxious olfactory stimuli to control and/or modify human behavior in this way is an attractive concept for modern warfare." Similar work is being done at Edgewood, with one smell called "U.S. Government Standard Bathroom Odor."

Despite their relatively benign if disgusting nature, it is not clear that malodorants qualify as legal weapons under the Chemical Weapons Convention. They are, after all, chemical weapons. The Edgewood project sought odors that are "not incapacitating or a sensory irritant," in that way avoiding classification of the malodorants as chemical weapons under the treaty. However, since a stink bomb is clearly intended to produce "sensory irritation," it would seem to fall within the same rubric as tear gas or pepper spray. If toxins produced by living things are used in the weapons, they might run afoul (so to speak) of the Biological and Toxin Weapons Convention as well, if not violating it then at least challenging its clarity, argues the Sunshine Project. If hostilities erupt, the group notes, malodorants could also be confused with more lethal chemical weapons, perhaps prompting an aggressive response that results in unintended escalation.

NO PAIN, NO GAIN

Since the mid-1990s, the U.S. Air Force and the Joint Non-Lethal Weapons Directorate, along with several private contractors, have been developing an "active denial" system (ADS) that causes pain but apparently no physical injury in individuals up to 770 yards away. The ADS basically sends a beam of microwaves (similar to those used in the kitchen ovens) that penetrate 1</>64 of an inch into the skin. It seems that a two-second burst can heat the skin to 130 degrees Fahrenheit. The idea is that a person will hastily get out of the beam to avoid the discomfort, moving away from the area that is deemed sensitive by authorities. Because of the low energy levels used, there is no real burning, only the sensation, unless the exposure is greater than 250 seconds, according to reports. However, there is reason to think that the cornea could be damaged much more quickly than the rest of the skin surface. It might be easy to protect against the beam with heavy clothing or other substantial barriers, limiting the ADS's effectiveness. Weather conditions such as rain might also limit the device's effectiveness by absorbing the waves.

Nevertheless, studies to mount the ADS on Humvees are fairly far along. In November 2004, the Raytheon Company delivered a prototype to the U.S. military for evaluation in 2005. According to the *Boston Business Journal,* the company has a four-year, $40 million development contract. The company CEO told investors that "this is where the future is going. This is the ability to protect our troops, and we're talking about the speed of light."

There are other options for creating what is intended to be nonlethal and noninjurious but debilitating pain, some yielded by basic medical research with interesting implications for weapons development. Research on nerve endings called nociceptors (noci- for "noxious") that react only to strong stimuli has been directed toward alleviating the effects of chronic pain. Unlike most receptors, nociceptors become more sensitive the more they are stimulated, so for those who suffer from persistent pain, gradually, less and less stimulation is required to set them off. In the 1990s, it was discovered that human platelets in blood plasma can themselves stimulate nociceptors when they release substances such as serotonin and histamine.

In spring 2005, the Sunshine Project obtained documents showing that the U.S. military is developing an NLW that takes advantage of this nociceptor reaction. A summary of the heavily redacted documents' content was published online by *New Scientist*. Called pulsed energy projectiles (PEPs), the devices take advantage of properties of plasma, ionized gas (not to be confused with blood plasma) in which the electrons are so highly energized that they escape from their atomic shells. When a pulsed laser hits a solid object, it can cause plasma to expand, producing an electromagnetic pulse that in turn activates nociceptors. PEPs are reported to have produced "pain and temporary paralysis" in animals, according to the U.S. Naval Studies Board. The magazine reported that work on intensifying the effect of PEPs without damaging tissue is being conducted at the University of Central Florida in Orlando.

HUMAN TESTING

At some point, it would seem desirable to test nonlethal weapons like the PEP device on human subjects, for reasons of safety as well as efficacy. Unlike more familiar ballistic or explosive devices, which can often be tested on inert objects or animals to get the desired information, many of the measures classified as NLWs involve human perception and behavior, so they must be tested on humans. Military powers, including the United States, have a long and unhappy history with this problem, including efforts to create an earlier generation of weapons not meant to kill. During the 1950s and 1960s, both the CIA and the Pentagon engaged in experiments with hallucinogens such as LSD and mescaline. There was interest in learning about the possibilities that these drugs could be used as a truth serum with spies or to sow confusion among fighting units. Hundreds of soldiers were given LSD, and many later complained of continuing emotional problems. In 1953, a patient hospitalized for depression died in an Army experiment to which he had not consented. Defense against lethal chemical weapons was also the subject of research. A British Royal Air Force engineer died in a sarin gas experiment at the UK Porton Down testing facility in 1953.

These cold war studies were officially justified as intended to develop defensive capabilities, as it was suspected that the Soviet Union was ahead of us in this field and might use such agents in an offensive capacity. A similar argument could, and surely will, be made today: that we need to know how these kinds of measures work so that we can protect our people against them. The trouble is that in the course of learning how to defend against a weapon, scientists also learn how to create it—which is very close to what happens in offensive weapons development. Just a week before the September 11, 2001, attacks, the CIA and Pentagon acknowledged to journalists that they were developing a strain of anthrax thought to have been developed by the Russians that resisted current medical treatment, and replicas of bomblets to disseminate chemical or biological agents. The purpose was defensive, but the knowledge gained could not be so easily limited to defense alone. There's no evidence that any human testing was done in these programs, but after 9/11 public tolerance for the idea of such testing undoubtedly increased.

Now, what would be the specific justification for human testing of the pulsed energy projectile? One reason would be the fact that pain is a somewhat subjective experience, and the behaviors of animals, even higher primates, might not be useful in setting dose limits. Also, pain experience is associated with brain receptor sites that could have lasting effects, perhaps triggering mental illness in some people. Again, exposures should be carefully established. The most obvious candidates for experimental subjects might seem to be people in uniform, but incidents such as the LSD experiments have led to rules that make it very difficult for people in the armed forces to be subjects. Before 9/11 and the massive redeployment of American forces, medical corpsmen and Special Forces personnel were used in highly regulated and supervised defensive biological and chemical experiments, but manpower shortages have greatly curtailed this approach. And there has long been ambivalence among military leaders about seeming to reduce their soldiers to "mere guinea pigs," even though the other risks to which they may be exposed are much greater.

Another traditional source of human experimental subjects is long-term prisoners. They were used during World War II in various studies, including a White House–sponsored malaria experiment in three federal penitentiaries that was described in national magazines. But after the war, prison experiments by the military were cast under the dark cloud of the Nazi concentration camp experiments. Fearing negative associations,

American security agencies started veering away from prisoners, though not before two died in a hepatitis experiment. Today, it is far more likely that any unclassified experiments would seek to enlist those whom scientists call "healthy, normal volunteers" through newspaper ads. Besides ensuring fully informed consent, those in charge would also have to decide how to set safety levels for people who have nothing to gain from the experiments (except perhaps a little cash for their time and inconvenience) and might be at least temporarily harmed.

In some cases, new countermeasures can be evaluated according to historic experience in other settings. For example, the antibiotic ciprofloxacin has been used successfully to treat cutaneous (skin transmitted) anthrax, and partly on that basis it has been approved for use in the treatment of inhalational anthrax as well, even though it hasn't been formally tested in human beings for anthrax infection conveyed through the air. The safety of these drugs or devices could be tested in humans, but their efficacy couldn't ethically be proven because it would be wrong to deliberately expose someone to anthrax for this purpose. So, instead, one can appeal to reasonably similar experience. Unfortunately, for many of the neuroscience-based approaches in this book there is either no historic experience (hypersonic sound is new), or the experience is so different that it makes analogies suspect (such as the use of anesthetic agents as crowd control measures). And animal experiments will not be useful for more subtle psychological effects.

These are not academic questions. The Air Force Research Laboratory's Active Denial System has already been tested on military and civilian employee volunteers, according to GlobalSecurity.org, which also reports that "the tests have been reviewed and approved by a formal Institutional Review Board [IRB] with oversight from the Air Force Surgeon General's Office." The tests appear to have been done in careful compliance with the federal regulations for doing human experiments. Reports assert that the volunteers provided informed consent and that the risks of the study were explained to them. The IRB found that the risk level was minimal, easing the way for using healthy people who could not benefit from the experiment, and the subjects were not paid for participating. Giving the studies more credibility, some of the project's own scientists volunteered for the experiment. As it turned out, the volunteers did experience only "minor skin tenderness due to repeated exposure to the beam," according to the Web site's account.

But what if the risks entailed more than the momentary pain or minor irritations from microwaves that were easily in the subject's control in the Active Denial System tests, but could cause potentially significant suffering or long-term health hazards? In that case, the research ethics committee might be less willing to approve a study that national security officials regard as important.

Cynics might point out that there is an easy way around the problem of experimentation ethics: just call the projects field trials instead of human experiments. That way, the medical ethics rules might be evaded. During the cold war, sailors and soldiers were exposed to the atomic bomb and chemical nerve agents. These incidents were categorized as field tests of environmental contaminants or training exercises rather than as human experiments. As a result, the rules then in place that required voluntary consent weren't technically relevant, even though in some cases soldiers wore radiation badges and had their bodily fluids checked. One can argue whether there was deliberate duplicity on the part of military officials in these cases or not, but I don't think so. In fact, it's often not easy to tell the difference between developing a new device or training for a new situation and conducting a human experiment. Sometimes it seems the key factor during the cold war era in determining whether an activity was a medical experiment was a straightforward one: If doctors were in charge, it was considered an experiment; if they weren't, it wasn't.

Besides the conduct of human experiments, NLWs raise other questions of professional medical ethics. A year after the Moscow theater tragedy, Robin Coupland of the International Committee of the Red Cross published a commentary in the *Lancet* in which he raised a number of questions for the medical community about its relationship to so-called nonlethal weapons, including whether health professionals should be trained to take care of people exposed to such weapons and whether they should allow themselves to be part of developing weapons that require medical and scientific expertise. Coupland noted that one medical research group suggested that international standards of conduct for NLWs be developed. It seems clear that eventually the international medical community is going to take up this issue.

NONLETHALS AND THE LAW

Since even so-called nonlethal weapons can injure and kill as well as produce grave suffering, how are national security policymakers guided in determining when an NLW is acceptable? The Pentagon has an internal review process for all weapons, including NLWs, using criteria that basically track international law: The weapon must not cause "unnecessary suffering" (a provision of the 1907 Hague Convention), must be capable of being controlled in a discriminatory manner (ruling out weapons that inevitably affect civilians as well as combatants), and must not violate a specific law that prohibits its use. But these criteria are very modest. Except for biological and chemical weapons of a patently offensive and deadly nature, experts disagree whether these rules and the international law they are based on pose significant obstacles to research, development, and deployment of very powerful NLWs.

Some authorities believe that current treaties prohibit nonlethals, but others note that many of these devices and agents could be used for defensive purposes and to protect innocents. David P. Fidler, an expert on international law at Indiana University Law School, told Jane's Information Group in 2000 that "if NLWs become more sophisticated and powerful, their potential may alter how experts look at the morality and legality of humanitarian intervention, anticipatory self-defence, enforcement of sanctions, and attacks on terrorist groups." He concluded that "the relationship between international law and NLWs will be more complex, controversial, and dangerous than people may realise."

Many chemical weapons are especially insidious and exemplify the inadequacy of current rules. They work by disrupting the nervous system's usual electrochemical pathways. Nonlethal chemical weapons would be different from deadly nerve gases such as sarin and soman. They would resemble more the old-fashioned psychochemicals such as LSD and BZ. The infamous Edgewood Arsenal LSD experiments in the 1960s showed that after less than twenty-four hours, the men who were drugged could fire nearly as accurately as those who were not. But BZ in small quantities is a calmative that causes sleepiness and loss of alertness fairly quickly, and after a few hours the individual can't react to surrounding events. In theory, BZ is a very attractive psychochemical weapon. It is also consistent with the rules that permit riot control agents to be used so long as their effects disappear in a few hours. In 2002, the *New York Times,* not known as a martial publication, editorialized that "in an age of terrorism, it would surely be desirable to develop a mist that could put people to sleep quickly without harming them permanently."

Yet defense officials still protest that there are substantial and unreasonable legal obstacles to developing and using nonlethal chemical weapons. Secretary of Defense Donald Rumsfeld has complained to Congress that international chemical weapons treaties make it acceptable to kill aggressors but not to immobilize them. Although this seems paradoxical, many of the chemicals being considered are based on toxins, and they don't always work as planned, as in the case of the Moscow theater incident. Since international law prohibits weapons that cause unnecessary suffering, any measure considered nonlethal would in theory also have to satisfy this criterion, vague though it is. For example, acoustic weapons like the hypersonic device would presumably satisfy this condition if they cause the target to lose balance and thereby become ineffective. But if they also cause targets pain that was not necessary to neutralize them, that would probably fall on the wrong side of unnecessary suffering. Calmatives that cause long-term disabling effects would also be prohibited under the Chemical Weapons Convention.

One plausible interpretation of the rules is that they prohibit the development of weapons *intended* to cause needless suffering, though of course accidents and unintended suffering can occur. Even with this generous interpretation, there are significant holes in the legal consensus. For instance, after the first uses of gas weapons during World War I, the 1925 Geneva Protocol found them "justly condemned by the general opinion of the civilized world." However, the United States has never interpreted this passage as prohibiting the use of, for example, tear gas for riot control. This common exception raises worries that new NLWs could be used as part of domestic police measures. Proliferation is another worry. The black market in intellectual and material resources needed to create atomic, biological, and chemical weapons is a matter of grave international concern. New classes of weapons, even if billed as not intended to kill or permanently injure, can significantly complicate arms control efforts and in some cases could be applied for lethal purposes for which they were not originally intended.

The Sunshine Project, the organization based in the United States and

Germany that tracks developments in new bioweapons, is especially concerned about the prospect that the spirit if not the letter of the Chemical Weapons Convention is being evaded. Psychoactive substances could be defined as riot control agents, and the unrest in which they are used could be defined as operations other than war. Referring to the Defense Science Board's 2004 recommendations to pursue NLWs, the Sunshine Project notes that the Joint Non-Lethal Weapons Directorate has classified research programs and has provided classified instruction to Marine Corps officers on nonlethal antipersonnel weapons. According to a statement by the director of the Sunshine Project's U.S. office, Edward Hammond, which is posted on the group's Web site, "We fear that JNLWD has new chemical weapons that are nearly ready for use, and that the DSB recommendations reflect another attempt to take JNLWD's chemical program out of the closet and put it on the battlefield."

JUST WAR

There is a reputable philosophical justification for developing NLWs: classical just war theory, which is often identified with St. Augustine but to which Cicero, Aquinas, and many others have contributed. The tribunal for war crimes in Nuremberg after World War II and the United Nations Charter have also added to just war doctrine. The criteria for a warranted war gained quite a bit of attention in the run-up to the invasion of Iraq, with philosophers and theologians considering whether in that framework the rationale for the Iraq war fell short.

Augustine argued that a war must be fought with the intent to attain peace and, apparently following the Roman philosopher and orator Cicero, that it must be undertaken under lawful authority. St. Thomas Aquinas added that the war must also be undertaken for a just cause. The Enlightenment Dutch Protestant thinker Grotius offered three criteria: the danger faced by the nation is immediate; the force used is necessary to adequately defend the nation's interests; and the use of force is proportionate to the threatened danger. As perhaps the founder of international law, Grotius placed importance on consultation among nations in his formulation of a just war. Critics of the George W. Bush administration's Iraq decision focused on what they contended was inadequate signoff by the

United Nations, which seems to have been required under Grotius' model. Interestingly, Secretary of State Daniel Webster explicitly adopted these rules on behalf of the United States in 1842.

Of Grotius' principles, that of proportionality is especially relevant to the ethics of NLWs. Of course, his principle is more about acceptable conduct in a war *(jus in bello)* than about the justification of war itself *(jus ad bellum)*. So, in theory, a party to a conflict may not use more force than necessary to achieve success. More specifically, one side is not entitled to inflict vast damage in response to a minor attack. Under proportionality, special care must be taken not to hurt noncombatants, seemingly ruling out any justification for collateral damage. Of course, one notable feature of the "total warfare" characteristic of many modern conflicts, starting with the American Civil War, has been that no part of the society is shielded from the effects of violence. From the standpoint of proportionality, Rumsfeld's complaint that the legal obstacles to developing NLWs are perverse has merit, as it does seem strange that chemicals that can kill are acceptable but chemicals that restrain are not. However, as we have seen, NLWs may not be as easy to control in practice as they are in theory. Careful case-by-case assessment is required to determine whether they fulfill their superficially attractive moral purpose.

8:

TOWARD AN ETHICS OF NEUROSECURITY

I USE THE TERM "NEUROSECURITY" to refer both to the ways that science and technology targeted at the brain and nervous system should be managed for the public good, and the means that democratic states must develop to protect themselves from their adversaries. As in the fields of biosecurity and atomic security, neurosecurity is complicated by the problem of dual use, the underlying and inescapable fact that medical and scientific breakthroughs can also be used for purposes unrelated to the goals of the researchers. The dual use problem is as old as human ingenuity itself. Even fire and the wheel were likely applied to intertribal conflicts as soon as the opportunity arose.

In a 2003 report on biotechnology and terrorism, the National Academy of Sciences, the nation's most prestigious scientific body, defined dual use technologies as "technologies intended for civilian application that can also be used for military purposes." The report went on to note that universities, private industry, and government labs are doing important experiments to find new treatments for AIDS, cancer, diabetes, and bacterial diseases as well as neurologic disorders such as Alzheimer's and stroke. Much of this research is directed to finding ways to detect harmful microbes and chemicals in the environment and to developing vaccines. However, "weaponization" of materials in labs is a concern that is taken much more seriously now than before 9/11 and the anthrax attacks, as it is feared that a rogue scientist might make off with a dangerous agent. This is one of the theories regarding the five anthrax deaths in the fall of 2001, but the crime has never been solved.

NEUROSCIENTISTS AND NEUROSECURITY

One theme I want to highlight in this chapter is the need for the scientific community to be more engaged in dealing with the unintended consequences of their work. Michael Moodie, the former director of the Chemical and Biological Arms Control Institute, has observed that "the attitudes of those working in the life sciences contrast sharply with the nuclear community. Physicists since the beginning of the nuclear age, including Albert Einstein, understood the dangers of atomic power, and the need to participate actively in managing these risks. The life sciences sectors lag in this regard. Many neglect thinking about the potential risks of their work." My experience suggests that an increased sense of the need to be publicly involved is taking hold among life scientists, especially in the face of recent controversies about stem cell research and intelligent design. Questions of dual use also require the informed engagement of our best scientific thinkers.

The dual use issue becomes more pressing as the science becomes more powerful and as more people possess the knowledge to apply it. We've seen that the applications of neuroscience and other brain-targeting fields to national security are no longer in the realms of science fiction or paranoid fantasy, that tremendous advances have been and are being made in understanding the way the brain works and, more slowly, in modifying it. Even though some of the claims that are being made are likely exaggerated, especially by companies trying to sell their products, not all are. Separating the wheat from the chaff is a challenge, but it does seem clear that the fascinating science I've described is on a course that, although not wholly predictable, almost surely points to greater understanding of and control over brain-related processes, and from various approaches.

When I've raised questions about the dual use implications of advances in brain science and technology at scientific meetings, many neuroscientists are surprised. Although they may receive Pentagon or CIA funding, brain scientists generally don't regard themselves as contributing to warfare. Those whose research is funded wholly by civilian agencies are taken aback when I suggest that their published results might well be examined by national security agencies to assess their implications. A number of neuroscientists have told me that they have received phone calls out of the blue from security officials interested in their work in areas such as monitoring or altering neural processes. Among those researchers who do accept national security agency funding, some tend to dismiss the idea that anything of military use will come of their research. Some believe, or prefer to believe, that they can manipulate their funding sources so that they can do the work they want to do without serving the goals of their benefactors, and that their results are going to be benign no matter what others might be looking for.

Often it's true that scientists are smart enough to get their grants without delivering the goods their funders want. But in the long run, as enough knowledge is gathered, the opportunities for dual use can't be completely avoided. For those who are deeply concerned about the exploitation of science for military purposes, an obvious answer seems to be that the scientific community should simply swear off cooperation with national security agencies, including accepting research contracts. Call this the purist approach. Based on some historical experience I shall elaborate, I believe the purist answer is shortsighted. In the real world, this kind of research is going to continue, and it's best that university researchers be those who do it, rather than building top secret science fortresses with researchers who are not answerable to anyone but their commanders. It is critical for the well-being of our democratic society that the civilian scientific community is kept in the loop and that the rest of us can have at least a general idea of the kind of work that is being done, even though for legitimate reasons many of the details may not be generally available.

An important reason to keep the scientific process as normal as possible, including transparency in interactions among scientists, is that science sets an example for an open society in which secrecy is minimized. Secrecy makes it harder for our elected representatives to fulfill their constitutional responsibility of overseeing government-funded science, and for experts outside of government to contribute to sound policymaking. One way a democratic society can minimize secrecy is to keep national security agencies linked to the larger world of academic science. For the same reason, suggestions in Congress and elsewhere that DARPA should pull back on its external funding should be resisted. The link between the academic world and the national security establishment makes for a healthier society than if each were isolated from the other.

There is of course another good reason that our best and brightest scientists should not reject this relationship: in a dangerous world, we do need to be protected. Though we should worry about the proliferation of new weapons and the diversion of resources necessary to develop them, the fact is that there are some rather malevolent forces out there. I am no pacifist and I do not advocate unilateral disarmament. While we need to be vigilant about dual use and the way our liberties may be affected, we also need to acknowledge the realities of modern threats to public safety. If our brain science and technology can provide us with some advantages, within carefully crafted constraints, it would be foolhardy, and perhaps immoral, not to explore them.

NO "NEUROPREPARAT"

Readers might be surprised that I reach this conclusion after cataloging the historic and modern examples of the not always admirable interest that national security agencies have shown in the brain and mind. Why wouldn't I advocate that our growing knowledge of the brain simply be made off-limits to military exploitation? For that matter, why not urge that no new science be studied for its potential in helping wage war? While admirable as an aspiration, in practice this position would lead to baleful consequences. In designing policy, we must acknowledge that neither science nor its martial applications can ever be static. While we might wish that no new kinds of weapons were added to the terrifying and massive arsenal already at the disposal of fallible human leaders, we need to find practical ways to address the problem. Similarly, rigorous separation of military research and civilian science will only result in war planners locating all research in facilities off-limits to public view (even more than is now the case), and creating a cadre of scientists who are beholden only to their government masters. Down that road lies the danger of powerful science fully captured by the state and beyond the reach of civilian control.

The history of science shows why we need to keep a close eye on the dual use problem. Though concern about dual use in biology is relatively recent, the atomic physicists of the 1940s and their intellectual descendants have been worrying over this problem since even before the atomic bombs were dropped at Hiroshima and Nagasaki. What lessons can dual use experiences in biology and physics provide as we face similar challenges in the applications of neuroscience?

Like the brain sciences, the development of modern microbiology is closely linked to the possibilities of warfare. The pioneering nineteenth-century microbiologists Louis Pasteur and Robert Koch developed techniques to isolate and culture microbes like anthrax so that medical defenses could be devised, but in short order their discoveries were applied to making biological weapons. In microbiology as in chemistry and atomic physics, the line between knowledge for defensive purposes and knowledge for offensive purposes is vanishingly thin.

Although the idea of using germs in warfare has been around since the ancient world—generally in crude forms like flinging the carcasses of infected horses over battlements—modern microbiology makes it feasible to select and even design agents in a "rational" rather than merely empirical manner. During World War I, Germany was accused of infecting horses of enemy forces with glanders, a highly contagious bacterial disease, and Japan engaged in what can only be described as an industrial effort to make massive quantities of bacteria in Harbin, Manchuria, during World War II, accompanied by horrific human experiments. At the same time, the United States, the United Kingdom, and Canada worked together in a secret offensive bioweapons program through the Army Chemical Warfare Service, though no deaths were intended or attributed to the Allies' efforts.

After the war, many bioagents were used in military and intelligence research until the United States ended its biological weapons program in the late 1960s. Much of the work was done at Fort Detrick in Frederick, Maryland, where treatments were devised for a number of infectious diseases with patriotic volunteer soldiers largely drawn from the Seventh-day Adventist Church. In 1972, the United States signed the international treaty on bioweapons, the Biological and Toxin Weapons Convention (BTWC), which made clear that only research for peaceful or defensive purposes is allowed. The Soviet Union was also a signatory, but didn't believe that the American program had really ended. In their suspicion, the Soviets developed a huge top secret system called Biopreparat that operated under civilian cover until at least 1992. Biopreparat continued to function even as Mikhail Gorbachev was deconstructing the country.

Ken Alibek was second in command of the former Soviet Union's biological weapons program until his defection to the United States in 1992. In his book *Biohazard,* Alibek describes the mammoth secret establishment put in place to counter suspected U.S. intentions in biowarfare development.

> Our program paralleled the Soviet nuclear program in organization and secrecy. Both generated a sprawl of clandestine cities, manufacturing plants, and research centers across the Soviet Union. The atomic weapons network controlled by the Ministry of Medium Machine Building was much larger, but the production of microbes doesn't require uranium mines or a massive work force. When our biological warfare program was operating at its peak level, in the late 1980s, more than sixty thousand people were engaged in research, testing, production, and equipment design throughout the country. This included some thirty thousand Biopreparat employees.

The Soviets in effect re-created the intellectual and material resources that were equivalent to several major research universities, but instead of using institutions that were integrated into the rest of the society, they created a huge clandestine system of science. Perhaps only an authoritarian system like that of the former Soviet Union could manage to sustain such a massive covert scientific establishment, or perhaps the creation of such a system would help lead to an authoritarian state. Either way, it is clear that the isolation of national security–related science from the larger scientific community is neither a viable nor a desirable option for an open society.

Moreover, Biopreparat's lack of accountability to Soviet society, repressed though that society was, arguably undermined Soviet quality control systems, resulting in threats to the public's well-being, such as the accidental release of anthrax from the biological weapons facility in Sverdlovsk in 1979. The exact death toll is unknown, but estimates range from sixty-six (the official Soviet figure) to over a hundred. A thorough KGB cover-up of the actual events—a defective filter on fermenting equipment was removed by a worker but never replaced—fooled even distinguished American scientists sent to investigate. Internally, military and government officials were in a conflicted position, as no one wanted to point fingers of blame. "The determination with which Soviet officials set about concealing the Sverdlovsk leak from their own people as well as the world was, under the circumstances, not surprising," Alibek wrote. "The truth would have severely embarrassed the nation's leaders, many of whom were not even aware that biological arms production was under way, and caused an international crisis." In representative democracies, both legislative oversight bodies and independent watchdog organizations play a significant role in keeping responsible parties accountable.

There are also security reasons to resist locating all potentially sensitive research in top secret institutions walled off from the rest of the scientific world. Secrecy will encourage states that feel threatened by our intentions to engage in exactly the sort of proliferation we don't want. Even the mere suspicion of secret activities is enough to encourage those who champion proliferation in their own systems, as was the case in the Soviet Union. Rather, we need to reassure others through the most transparent verification programs possible. Other countries may also use our secret neuroscience as an excuse to cloak their own programs. And if our superior defensive capabilities are known to potential adversaries, they will be less likely to be interested in probing for weaknesses. Progress in neuroimaging techniques that provide evidence that an individual is familiar with a certain place (a terrorist training camp, for example) could prove helpful in distinguishing between more and less likely sources of information among captives. Similarly, many scientists argue that the best way to ensure there are defenses that can neutralize new weapons of any kind is to allow the scientific community at large to learn about them—within reasonable limits of course—as scientists can then also learn what their weaknesses are and what countermeasures can work. Secrecy about science is not necessarily good for our security. It can have exactly the opposite effect.

The shadow of Biopreparat survives. Fears persist that some of the vicious new biological agents that were being attempted in those Soviet-era labs may be stored somewhere, in spite of vigorous attempts to identify and clean up the sites. Certainly the expertise that was created is stored in the heads of many ex-Soviet scientists, and the United States and other countries have made strenuous and apparently successful efforts to ensure that this two-edged sword of scientific knowledge is not hired out to the highest bidder.

LESSONS FROM ATOMIC PHYSICS?

My view that it would be bad for science and for our society for the neuroscience community to insulate itself from support from national security agencies does not imply that research should be unconstrained. Rather, I believe that the neuroscience community needs to be part of discussions about the conditions for entering into relationships with security agencies and the guidelines that would govern the research. Many of these rules, such as the standards that govern experiments involving human subjects, are already in place but are of a technical nature. There are larger philosophical questions about the social obligations of scientists.

The most dramatic case of a scientific community grappling with its moral responsibilities regarding military applications of its work is that of the atomic physicists and nuclear weapons. Although differences of opinion appeared from the very earliest days of work on the superweapons, for many physicists the turning point came after the development of the hydrogen bomb, a weapon of far greater destructive capacity than the original atomic weapons. A number of them turned decisively toward the antiwar activists and struck up an alliance. The leader of that group was the British philosopher Bertrand Russell, a longtime socialist. Einstein decided that one of his last acts would be to join his prestige with that of Russell, and together they drafted a statement known as the Russell-Einstein Manifesto. Published on July 9, 1955, the manifesto expressed the gloom that had descended upon many of the original Manhattan Project scientists:

> No doubt in an H-bomb war great cities would be obliterated. But this is one of the minor disasters that would have to be faced. If everybody in London, New York, and Moscow were exterminated, the world might, in the course of a few centuries, recover from the blow. But we now know, especially since the Bikini test, that nuclear bombs can gradually spread destruction over a very much wider area than had been supposed. . . .
> There lies before us, if we choose, continual progress in happiness, knowl-

edge, and wisdom. Shall we, instead, choose death, because we cannot forget our quarrels? We appeal as human beings to human beings: Remember your humanity, and forget the rest. If you can do so, the way lies open to a new Paradise; if you cannot, there lies before you the risk of universal death.

The manifesto resolved that only the abandonment of war by governments, and especially by the United States and the Soviet Union, could rescue humanity from imminent catastrophe. Although war has surely not been renounced, the Russell-Einstein statement did catalyze an international anti–nuclear weapons movement that arguably helped create the test-ban treaties of the 1960s and the drawdown of armed missiles in the 1980s. Gradually, the atomic physics community shifted its position to a "no first use" philosophy, which seemed a far more achievable constraint to which all nuclear nations could, at least rhetorically, commit themselves. The manifesto and the movement it stimulated showed that scientists, especially Nobel laureates, could employ their prestige to influence the political use of their science.

LIMITS OF THE BOMB ANALOGY

Unfortunately, "no first use" doesn't fit well in the context of neuroscience and national security. There are a number of important differences between the atomic physicists' experience and that of the neuroscientists; most neuroscience is aimed at either healing directly or understanding the brain well enough to do so. But the original atomic physicists, after some initial uncertainties about the technical possibilities of nuclear fission, knew that they were developing a weapon of unprecedented destructive capacity. It was their intention to do just that, out of fear that the Germans would get there first. In fact, the war effort's funding and support for the Manhattan Project paved the way for the knowledge and technologies that made possible the peaceful uses of atomic energy, such as reactors to produce electricity. One could almost say that atomic energy was dual use turned on its head, as military research and development created the conditions for the civilian use.

Another important difference between nuclear and neuroscience weapons is that, as became clear to war planners in the 1950s, especially after development of the hydrogen bomb, nuclear weapons are muscle-bound. Although "improvements" have been made that modify the nuclear bomb's fission release, in general the bomb is too powerful and its effects too uncontrollable to provide tactical advantage on a battlefield. What nuclear bombs mainly afford their owners is strategic advantage by influencing the behavior of potential adversaries. Thus "nuclear blackmail" and "mutual assured destruction" became familiar concepts during the cold war. The utility of neuroscience-based weapons, however, is mainly tactical, in that they might provide short-term and relatively targeted advantages such as disrupting an enemy patrol or disabling a terror cell. In that sense neuroweapons are much more manageable than nuclear weapons. Not weapons of mass destruction, they are better considered weapons of selective deception and manipulation.

Yet another difference is that it's much easier to distinguish between the offensive and defensive use of atomic bombs than offensive and defensive applications inspired by or based on neuroscience. As I've noted, innovations that focus on the brain and nervous system may be applied to enhancing the training, selection, and prospects for survival of troops long before they are deployed, if ever. An imposing advantage could discourage adversaries or at least create various advantages in small confrontations. And, according to some, the atomic weapons industry creates harrowing environmental problems that have caused critics to conclude that the bomb is a curse with few long-term benefits for humanity.

Finally, the neurosciences and related fields may well lead to measures that both give us an advantage over our adversaries and are morally superior to other tactics, a combination of considerations that doesn't so easily apply to nuclear weapons. An example is interrogation. Naively, torture might seem like the easiest way to get someone to talk. But a brain bombarded with painful stimuli is going to have activated neural systems associated with fear and survival rather than cognition. I spoke about the shortcoming of violent interrogation techniques with Michael Grodin, a psychiatrist and bioethicist at Boston University who has worked with torture victims for years. "I've got experience with eight hundred torture victims and no one has ever been able to show that any critically useful information comes from torture," he told me. "And the more severe the torture, the more problematic the so-called information from the victim."

Scientifically informed techniques and devices seem likely to provide far more subtle ways to obtain information that is not polluted by the stress of talking under torture. Functional MRI studies of people who are playing a trust game with money have found increased activity in the caudate nucleus as one player learned to trust the other to invest his money. These signals appeared more quickly in the brain as the game went on and one player gained more confidence in the other. The caudate nucleus is linked to the brain's reward pathways so that it is more activated when there is an expectation of a positive event, such as being given some juice or, in this case, the socially rewarding experience of feeling good about someone else. If these signals could not only be monitored during interrogation procedures but the relevant pathways deliberately stimulated—setting aside for the moment ethical and legal concerns about peering into neural processes—it could advance the conditions for fruitful interrogation.

All in all, it doesn't look as if neuroscientists can take the same position about the often unintended fruits of their labor as the atomic physicists, that a "no first use" policy should prevail and they should only be used for defensive purposes. That policy makes sense for weapons of strategic value that might stimulate an uncontrolled exchange of weapons of mass destruction such as nuclear bombs, but it doesn't seem to work for tactical weapons that might head off more violence by, say, compromising a terrorist group holding civilian hostages. Nor is it clear how to apply such a policy to devices used in screening or training soldiers, or pharmaceuticals designed to improve a fighter pilot's cognitive capabilities. It seems that grappling with the ethical issues raised by the applications of neuroscience to national security will require an entirely different approach from that taken by the atomic scientists.

LESSONS FROM BIODEFENSE?

The field of biodefense is spawning an ethical debate of its own, and some of that work is relevant to neuroscience. For example, Ronald Atlas, a professor of biology and biosecurity expert at the University of Louisville, has developed a code of ethics for biodefense research. His code requires scientists to avoid doing anything to facilitate bioterrorism and obligates them to call the public's attention to any such activities, to restrict access to information that could lead to dual use, to ensure that the ben-

efits of the research outweighs the risks, to respect the rights of human research subjects, and to respect the conscientious objections of those who
decline to participate in the research.

These are admirable standards, and several apply directly to neurodefense research. One important practical difference is that research involving bioagents often results in potentially dangerous materials being kept
in laboratories, so the physical security of those labs poses concerns that
are not so common in neuroscience. But the emphasis on the social responsibilities of researchers, the obligation to consider risks and benefits
and to protect the rights of those who might be used in experiments, is a
useful precedent for neuroscience and national security.

The rules governing the biological weapons field present some interesting dilemmas when applied to neuroscience weapons. For instance, the
Biological and Toxin Weapons Convention prohibits the use of bioweapons to manage civil disturbances such as riots. It makes sense not to allow
law enforcement agencies to use anthrax, but perhaps not to ban the use
of fMRI for "lie detection" or even hypersonic sound to root out terrorists,
for instance. Whether these are acceptable remains to be debated, but at
least they raise different issues than bioweapons.

Potentially undermining any measures intended to police biodefense
research and development is the intensified secrecy surrounding biodefense research. Again, this experience should serve as a warning in the area
of neurodefense. The Federation of American Scientists' bioweapons expert, Barbara Hatch Rosenberg, has written in the journal *Disarmament
Diplomacy* that for about the first fifteen or twenty years after the Biological
and Toxin Weapons Convention came into effect in 1975, the Defense Department kept its program unclassified, except for results that could reveal
"U.S. military deficiencies, vulnerabilities, or significant breakthroughs in
technology." After that, perhaps as a result of the Gulf War and the revelations about the Soviets' Biopreparat, the policy seemed to change. Just one
week before 9/11, the *New York Times* reported on three secret biodefense
projects that strain the treaty. One was a plan to create a vaccine-resistant
anthrax strain that it is thought the Soviets had produced, ostensibly to
learn how to defend against it. Although U.S. Western allies were disturbed
by these revelations, they muted their reactions due to the 9/11 attacks.

The secrecy problem returns us to the need to retain connections between civilian science and the national security establishment. How can citizens make judgments about the policies their leaders are carrying out on their behalf if the information they have is so limited? It can't be assumed that universities are fonts of openness, however. They often participate in private financing arrangements that protect intellectual property. Universities also have differing policies for handling classified research. These policies should be the subject of public discussion and standardization so that the academic world does its part by requiring the greatest possible transparency.

For some kinds of science, greater use should be made of inspection regimes, but recent political decisions have hampered these arrangements. As Rosenberg writes in her article, "Secrecy is particularly corrosive, especially when combined with rejection of international monitoring. Suspicions would be largely dispelled if threat assessment projects were openly declared and subject to international inspection. There would be no need to disclose project results that impinge on national security." Obviously, finding the balance between national security and democratic openness is easier said than done, but the right start seems to be to make transparency rather than opacity the default position.

ETHICALLY REGULATING NEURODEFENSE

There's no easy fix to these issues, no bumper sticker solutions. They will have to be considered carefully and dispassionately by neuroscientists, agency officials, and representatives of the public. Many government advisory committees are already structured to provide and receive input from various sources. Rather than sweeping policies, the diversity of the neuroscience and its applications will challenge our ability to craft policies that tie familiar ethical concepts such as respect for personal autonomy to specific neuroscience-based techniques.

The mechanism for many of the decisions about the appropriate research and use of brain-based national security measures is already in place. The Department of Defense and the CIA are included in a federal regulatory framework called the Common Rule designed to protect human research participants; it requires review of proposals by a research ethics committee and the informed consent of the volunteers. The FDA regulates the licensing of drugs and devices. Though this system is far from perfect, it at least creates obstacles and some measure of accountability for the dissemination of much of the technology I've described.

But, clearly, the current regulatory system doesn't automatically answer all the questions about proposals for experiments with new drugs and devices or applications for licensure that might come before it. In some instances the technology is already licensed and can simply be applied to novel uses without going through the FDA process, as in the case of calmative drugs licensed as anesthesia for surgery. However, if experiments are to be done to see how well they work for some new purpose, such as managing a hostage situation, then informed consent and prior review are generally still required. All this could be bypassed if military authorities request a waiver of informed consent on national security grounds, as happened during the first Gulf War when agents thought to be protective against nerve gas and biological weapons were offered to troops even though the agents had not been approved for that purpose. Some policies and procedures will need to be in place for those experiments and applications of the new neuroscience and related fields that can't be captured in the routine regulatory process because of national security needs.

Often, the variety and potential usefulness for national security of innovations intended to affect the brain will require close ethical analysis. I asked Bill Casebeer, the terrorism expert and neuroethicist, how decisions are going to have to be made. His answer was partly framed in the technical language of ethics:

> Consideration of virtue theoretic [what are the moral potentials in the situation], deontic [what duties are at stake and to whom], and utilitarian [likely empirical consequences] aspects of the problem will be a useful starting point; I'd be surprised if there's anything that's entirely new in the field of military ethics that would be posed by consideration of neuroscientifically informed military operations. Considerations of virtue and vice, human functionality, rights and duties, consent, innocence, involvement in the causal and logical chain of agency required to do harm to another, outcomes in both the act and rule sense, etc., will all be operative. Some emerging technologies may make these considerations more pointed (e.g., does marketing informed by neuroscience somehow diminish the agency of those involved in marketplace transactions?), but the "entirely new" issues will be few and far between, I suspect.

I agree with Casebeer that in dealing with emerging neuroethical dilemmas in the national security context we can learn from previous ethical quandaries, especially in terms of the conditions under which we have those discussions. I also agree that in many cases, the ethically acceptable course of action will be a matter of weighing and balancing rather than appeal to an overarching moral doctrine, though basic guidance from some principles is going to be needed. For instance, a number of the scientists, lawyers, ethicists, and advocates with whom I spoke in the course of writing this book agreed that there had to be vigorous protection of at least one nonnegotiable premise when considering the appropriate security applications of neuroscience. In the law, this principle might be expressed in terms of the protections afforded in the Fifth Amendment of the Constitution regarding self-incrimination: "to be a witness against himself." Philosophically, this can be expressed as the proposition that no one else should be able to decide what goes into my brain or who "reads" it.

Like any philosophical principle, this one admits exceptions if they can be justified. One exception might be the example frequently cited by proponents of at least some limited torture option: the terrorist in our custody who is aware of a ticking bomb that could kill and injure many civilians. In that sort of case, advanced neuroscience-based technology might be helpful and its application justified. But justifiable exceptions to principles do not undermine their general validity. Suppose we can agree that, in any national security question regarding the brain, the presumption is that cognitive liberty is guaranteed in the absence of an overwhelming counterbalancing argument. The difficulty, of course, is that the need for the sovereign state to defend itself can easily be used as a trump card by legitimate political authorities. Under those circumstances, we have to rely upon some legal process to constrain state power. Again, the maximum possible transparency and accountability will have to apply.

One idea is to create the neurosecurity equivalent to the National Science Advisory Board for Biosecurity that was established in 2004. This new board is administered by the National Institutes of Health, but advises all cabinet departments, including the Defense Department, and "others as appropriate." Its mission is "to provide advice to federal departments and agencies on ways to minimize the possibility that knowledge and technologies emanating from vitally important biological research will be misused to threaten public health or national security." Focused particularly on the problem of dual use of biological agents, its charge includes developing guidelines for research and professional codes for scientists. The government also needs advice on what the most likely biological threats are and what sorts of countermeasures should be developed. Another panel, the Committee on Biodefense Analysis and Countermeasures, on which I serve, was created by the National Academies in response to a request by the Department of Homeland Security.

An advisory committee on neurosecurity could cover analogous problems of dual use and the implications of countermeasure development. If it was created by the National Academies, it could draw on talent from a range of disciplines. Diversity of scientific input is crucial for addressing national security issues. Scientists tend to work in silos, concentrating on their own discipline or subdiscipline or target of interest such as a certain organ or system, gene, or protein. Normally, this focused way of working is productive, but when novel real-world problems emerge, it can be overly limiting. National Academies committees are able to overcome these disciplinary boundaries.

An example of the need to apply several disciplines to neurosecurity problems is the way that genetically engineered biological weapons can become very scary neuroweapons. Bioweapons such as viruses have a payload: the genetic content of the virus; a delivery system, the outer viral coat; and a target, such as an organ system of the human body. All three components of the weapon system can be manipulated by pathogen genetic engineering. For example, certain viral and bacterial pathogens can be engineered by insertion of foreign or synthetic genes with properties not naturally found in the virus or bacterium to become advanced neuroweapons targeting the brain and nervous system. Based on work already done in the offensive biological weapons program of the former Soviet Union, scientists who are expert in biological weapons defense have worried aloud to me about the threat of technological surprise posed by advanced viral neuroweapons carrying synthetic genes coding for short peptides (short strings of biologically active amino acids with biological activity) into the central nervous system.

Inside the central nervous system, the technological surprise stems from designer peptides produced from synthetic genes that have effects quite distinct from those normally associated with the pathogen. For example, when produced in the brain, they could function as malign neuro-modulators, disabling brain functions by modifying the relationships and communications between neurons. In such advanced neuroweapons, the infectious pathogen is really just a Trojan horse, selected for its ability to get the synthetic gene quickly into a target it cannot otherwise reach.

The advanced neuroweapon does not necessarily have to enter the brain and nervous system to modulate function. An example is *Francisella tularensis* genetically modified to produce beta-endorphin. This bacterium, a well-known biological weapon in its native form, is the cause of tularemia, also known as rabbit fever because it's often found in rodents and can be passed to humans by direct contact or by inhaling the particles. Normally, tularemia can easily be treated with antibiotics, but if the bacterium has been engineered to generate a potent neurochemical, the damage would already have been done before the infection became a clinical problem. Various kinds of disabling reactions, from intense fatigue or confusion to the loss of sensation, could be attempted that would neutralize enemy forces.

I talked at length about this problem with a twenty-year U.S. biodefense expert who preferred not to be identified by name. He argues that rapid-onset, brain-targeted biological weapons are something we need to worry about now, not sometime in the future. "There is no point on the battlefield in exposing the opposing troops to a synthetic gene that is going to give them liver cancer in fifteen years or make them incapacitated next week," he told me over a pleasant lunch in Rosslyn, Virginia.

> There must be an immediate purpose, such as to disrupt the characteristics that enable men to protect themselves and to fight as an organized force, not a rabble with weapons. There is some interesting history here. Before the French Revolution, soldiers fought for pay or because they were given lands and had a vested personal interest. Astonishingly, and for the first time, the French Revolutionary Armies fought enthusiastically and very successfully for abstract concepts—liberty, equality, and fraternity. Clausewitz and others studied this phenomenon, and a whole body of doctrine evolved about what it takes to employ a citizen army of millions of enlisted or drafted soldiers who are fighting for patriotism and loyalty with no real personal gain or stake. The ability of units to function is all in the mind. So if one can disrupt unit loyalty

through fear or another emotion, the army would cease to exist as a fighting force. Claustrophobia would make soldiers tear off their protective face mask. Fear, thirst, accelerated heart rate, hypermotility of the gut—these would be the desired peptide effects.

Clearly, any attempts to engineer neuroweapons would be a violation of international law governing biologic and toxin warfare. But even with an inspection regime, there are no guarantees that the treaties will be honored. Reports that the Soviet Union was working on tularemia and viruses as candidate neuroweapons demonstrate that the genie is already out of the box. "A literal handful of people have been thinking about these issues for a decade or more, including in neuroscience," the expert told me, "but these are not the people who are now 'biosecurity experts' with lots of funds! We have got to think about these pathogens as weapons delivery systems and the critical scientific inputs we need are not those of infectious disease physicians but of those who can anticipate what might be inside the Trojan horse before it opens—because very soon after it does open, as in Troy, it might be impossible to mount a defense."

A national science advisory board for neurosecurity should have the clout to put all these neuroweapons issues on the front burner by combining a practical orientation with diverse scientific, legal, and ethical expertise. Composed of the most highly regarded scientists, its mission could be to advise all federal agencies that fund, apply, or regulate research that affects the brain and nervous system and could be used or misused for national security purposes. To give the problems the attention they deserve, perhaps this board would report to the National Security Council and review all the portfolios of those agencies engaged in neuroscience research. That would admittedly be a huge task, so the board might deal only with specific cases referred to it. In either case, such an expert board would perhaps be concerned both with monitoring dual use and helping develop policies on the applications of technologies that are targeted to the brain and nervous system in a national security context. The committee could also advise on the development of countermeasures. Considering the great variety of possible technologies to be covered, the membership would need to have correspondingly diverse expertise, including neuroethics.

A ROLE FOR NEUROETHICS

A government advisory committee on neurosecurity could benefit from including people who spend their lives trying to think clearly about ethical issues in science. Some of the best thinkers in both neuroscience and neuroethics have told me they would be interested in helping security agencies look ahead and develop policies. There are some early examples of members of the bioethics community being called upon to work on these problems by defense agencies and their contractors. Granted, you don't have to be much of a cynic to suspect that the agencies' interest in ethics is all show with no substance, but it seems to me there's more to it than that.

For instance, I am fairly confident that my friend Laurie Zoloth, a Northwestern University professor, is so far the only bioethicist to have been asked to give a lecture to DARPA. That the agency would care about ethics enough to listen to an academic presentation surprised no one more than Zoloth, who is a self-described left-wing peacenik bred in the sixties. Her biography includes dropping out of college to help save the world and working as a community organizer. Today she is an expert on Jewish moral philosophy as well as the ethics of science. A colleague on another project asked her to attend a DARPA retreat in 2002. The meeting included people from veterinary and medical schools from around the country as well as agency staff.

Zoloth gave her talk, with no honorarium, and was so fascinated by the lucid and serious discussion about the ethical theories she presented that she decided to stick around to hear the science lectures, which included presentations on improving the safety and protection of troops, wound healing, ways to better deliver health care in an extreme situation, self-care when injured, and so forth. She was especially struck that the agency officials appreciated that the implications of their efforts to make health care more deliverable when there are no health care professionals and few resources could as easily be applied to people in impoverished countries as to isolated soldiers, though that wasn't their immediate intent. Overall, Zoloth came away impressed, as she told me, by "their intelligence, their optimism and creativity, and their positive reception to thinking about ethics." She didn't expect to like them as she did, and later took some predictable criticism from colleagues who thought she was selling out to the masters of war.

As if one such experience wasn't enough, the DARPA chiefs invited her a second time (again she didn't accept compensation), on this occasion to a conference at their innocuous headquarters in suburban Virginia. Zoloth's topic this time was ethical issues that arise when technologies such as computation, nanotech, and genetics converge, leading to neural implants and worries about mind control. Again, the hundred-plus attendees raised fascinating questions about how we think about science as compared with the way we think about ethics, about cultural differences and their role in morality, even about theories of justice and access to scarce resources. Again, Zoloth found herself impressed. They "obviously weren't in it for the money," she reflected. "They were thoughtful about their power, open to new ideas, and in some ways were a refreshing change from many left-wing academic audiences, more open. They were patriotic, sure, but in a nonjingoistic way I found heartening."

DARPA and its counterparts aren't about to become departments of moral philosophy or Franciscan monasteries. But targeted ethical analysis on specific issues could and should help guide policy on acceptable areas of research. An example of the potential role of ethical analysis in the world of national security neuroscience was a proposed presentation at the first AugCog International Conference in Las Vegas in July 2005, held in conjunction with the Eleventh International Conference on Human-Computer Interaction. This was a sort of expo of applied neuroscience, with themes and topics such as "human performance engineering," "human-computer interaction," "engineering psychology and cognitive ergonomics," "multisensory interfaces," and just about every other bit of jargon that could have made its appearance in an Isaac Asimov novel. Once again, the work being presented was admirable, including numerous papers on developing assistive devices for people with disabilities. The range of work represented was staggering, a potent exhibition of the potential for sophisticated, practical neuroscience to change the way we deal with our world and the way we see ourselves in it.

How interested were the AugCog attendees in the ethical issues associated with their work? It's hard to know, but one piece of information is not encouraging. Two bioethicists, Judy Illes from Stanford and Paul

Wolpe from the University of Pennsylvania, were scheduled to be on a panel called "Exploring Ethics in Augmented Cognition Research," in which they offered to begin a dialogue between AugCog and ethics, two fields that they believe need to understand each other better. Both are exceptionally well qualified. Wolpe is a sociologist who works in both bioethics and psychiatry and is also NASA's bioethics adviser; Illes is recognized as one of the founders of neuroethics and has been asked by several DARPA-funded companies to assist with ethical issues. At the AugCog convention, Wolpe was to discuss the history and ethics of the use of human volunteers in research, Illes the particular issues that arise in neuroscience research. Unfortunately, the panel was canceled after only three people signed up for the session. "We really wanted to get this exchange of views on the ethical questions going," Illes told me, adding that "the apparent lack of interest in ethical concerns by DARPA-funded scientists is a challenge."

Is it pie-in-sky idealism to expect that entrepreneurs who make their living as defense contractors will take a nanosecond to worry about neuroethics? I might have thought so, but then I was introduced to Don DuRousseau, a veteran who owns a small company called Human Bionics LLC. DuRousseau had been talking with the Pentagon about developing a brain-machine interface system that puts together several of the concepts I've talked about in earlier chapters: a handheld device that wirelessly records and analyzes all sorts of biological data in real time for measuring the general health of the brain and body and how these systems respond to target stimuli. Such capabilities would let the computer side of the brain-machine system predict whether a soldier was losing a high level of alertness on a particular task and, if so, redirect his or her attentional resources to the significant environmental elements and improve arousal level. Thus, in addition to simple heart and respiration rate information, twenty-first-century brain-machine technologies will include ongoing assessment of working memory and attentional cognitive systems to improve overall job performance, train individuals and teams faster, and even extend cognitive capacities through external means.

The concept looks impressive, and DuRousseau has some top computer scientists working with him. Of particular interest to me is the fact that he has had ethics advice from the very beginning. His start as a neuroscientist working on basic research questions provided the rigor of the scientific method and exposure to the process of peer review and oversight by coworkers and seniors, he told me. This environment laid the groundwork for the way he views the importance of immediately and clearly establishing the ethical principles on which his company operates, particularly with respect to developing new products that expose aspects of the human side of the brain-machine interface. DuRousseau says that he's always been interested in learning about the brain: the source of our mind and human uniqueness.

> Of all the organs in our bodies, it's our brains that separate us from the apes and from each other, and those thoughts have been the driving influence behind my business. That being said, it places my business at the heart of passionate debate, not only over issues of our origins but also of the technologies now able to peer in at our uniqueness. Without a strong ethical foundation with clear operating policies, my business would fail as soon as it became known to the public that we can run machines with only our brains. No matter what the application, someone will find it unethical, so I have to be able to define the greatest good from my technology and weigh that against the costs of defending the ethics of my business.

CODA: BEYOND MIND WARS

Dual use is a two-way street. In this book I have primarily considered the military applications of neuroscience and brain-related technologies. But cooperation is as much a part of the human condition as conflict and its material manifestation constitutes the bulk of the evolved human brain. As with all successful species, our ancestors survived because they spent more time helping than hurting one another, which made it possible for the human cerebrum to find ways to thrive. We should be able to learn and apply the lessons of the new brain science for peaceful purposes. As the national security implications of neuroscience become more apparent, the pressing need to examine how our brains dispose us to peace as well as war should gain currency. The fields of conflict resolution and peace studies could enrich and be enriched by information from the neurosciences. Future interventions into international and civil conflicts may benefit from greater sophistication about the human brain.

The long-term trajectory of humanity combines a growing capacity for indiscriminate destruction along with vast increases in constructive methods and techniques for solving problems that inhibit human flourishing. Somehow, these seemingly contradictory traits must be neurologically linked. Perhaps, understanding more about this excruciatingly complex system, we can turn ourselves from the wars of the mind to the peace of the soul.

SOURCES

CHAPTER 1

Casebeer, William D., and James A. Russell. "Storytelling and Terrorism: Towards a Comprehensive 'Counter-Narrative Strategy.'" *Strategic Insights* 4, no. 3 (March 2005): 15–31.

Defense Advanced Research Projects Agency. Research and Development Presolicitation Notice, February 8, 2006. http://www.darpa.mil/baa/baa05-19pt1.html (accessed March 19, 2006).

Faden, R. R., S. E. Lederer, and J. D. Moreno. "U.S. Medical Researchers, the Nuremberg Doctors Trial, and the Nuremberg Code: A Review of Findings of the Advisory Committee on Human Radiation Experiments." *Journal of the American Medical Association* 276, no. 20 (1996): 1667–71.

Foerstel, Herbert N. *Secret Science: Federal Control of American Science and Technology.* Westport, CT: Praeger, 1993.

Goldblatt, Michael. "DARPA's New Research Frontiers," May 9, 2002. http://sdstc.ucsd.edu/Transcripts/goldblatt_T050902.pdf (accessed March 18, 2006).

Grove, J. W. *In Defence of Science: Science, Technology, and Politics in Modern Society.* Toronto: University of Toronto Press, 1989.

Hogan, Michael J. *A Cross of Iron: Harry Truman and the Origins of the National Security State, 1945–1954.* Cambridge: Cambridge University Press, 1998.

House Armed Services Committee. "Statement by Dr. Tony Tether, Director Defense Advanced Research Projects Agency, Before the Subcommittee on Terrorism, Unconventional Threats, and Capabilities House Armed Service Committee, United States House of Representatives," March 27, 2003. http://www.securitymanagement.com/library/DARPA_tether0603.pdf (accessed April 19, 2006).

India Daily Technology Team. "New Advanced Mind Control Techniques Give a New Flavor to Modern Warfare." *India Daily,* May 20, 2005. http://www.indiadaily.com/editorial/2799.asp (accessed July 31, 2005).

Korn, James H. *Illusions of Reality: A History of Deception in Social Psychology.* Albany: State University of New York Press, 1997.

Moreno, Jonathan D. "Human Experiments and National Security: The Need to Clarify Policy." *Cambridge Quarterly of Healthcare Ethics* 12, no. 2 (2003): 192–95.

———, ed. *In the Wake of Terror: Medicine and Morality in a Time of Crisis.* Cambridge, MA: MIT Press, 2003.

———. "A New World Order for Human Experiments." *Accountability in Research* 10 (2003): 47–56.

———. "'The Only Feasible Means': The Pentagon's Ambivalent Relationship with the Nuremberg Code." In *Bioethics, Justice, and Health Care,* edited by Wanda Teays and Laura M. Purdy. Belmont, CA: Wadsworth, 2001.

———. "Reassessing the Influence of the Nuremberg Code on American Medical Ethics." *Journal of Contemporary Health Law and Policy* 13 (1997): 347–60.

———. "Remember Saddam's Human Guinea Pigs." *American Journal of Bioethics* 3, no. 3 (2003): W53.

———. *Undue Risk: Secret State Experiments on Humans.* New York: Routledge, 2001.

Moreno, Jonathan D., and Susan E. Lederer. "Revising the History of Cold War Research Ethics." *Kennedy Institute of Ethics Journal* 6, no. 3 (1996): 223–37.

Yergin, Daniel. *Shattered Peace: The Origins of the Cold War.* New York: Penguin, 1977.

Weldes, Jutta, Mark Laffey, Hugh Gusterson, and Raymond Duvall, eds. *Cultures of Insecurity: States, Communities, and the Production of Danger.* Minneapolis: University of Minnesota Press, 1999.

CHAPTER 2

Arndt, Michael. "Rewiring the Body." *Business Week,* March 7, 2005. http://www.businessweek.com/magazine/content/05_10/b3923001_mz001.htm (accessed August 1, 2005).

Bloom, Floyd, Charles A. Nelson, and Arlyne Lazerson. *Brain, Mind, and Behavior.* 3rd ed. New York: Worth Publishers, 2001.

Defense Advanced Research Projects Agency. Information Processing Technology Office. "Cognitive Information Processing Technology Proposer Information Pamphlet." http://www.darpa.mil/ipto/solicitations/open/02_21_PIP.htm (accessed March 21, 2006).

———. "Improving Warfighter Information Intake Under Stress." http://www.darpa.mil/ipto/programs/augcog/ (accessed January 28, 2005).

Goldblatt, Michael. "DARPA's New Research Frontiers," May 9, 2002. http://sdstc.ucsd.edu/Transcripts/goldblatt_T050902.pdf (accessed March 18, 2006).

Hoag, Hannah. "Neuroengineering: Remote Control." *Nature* 423 (June 19, 2003): 796–98.

Hoffman, Robert R., Jeffrey M. Bradshaw, Patrick J. Hayes, and Kenneth M. Ford.

"The Borg Hypothesis." *IEEE Intelligent Systems* 18, no. 5 (September–October 2003): 73–75. http://www.computer.org/intelligent/homepage/03x5hcc.htm (accessed December 23, 2004).

Huang, Gregory T. "Mind-Machine Merger." *Technology Review* 106, no. 4 (May 2003): 38–46.

Kageyama, Yuri. "Remote Control Device 'Controls' Humans." Associated Press, October 27, 2005.

Marriott, Michel. "Robo-Legs." *New York Times,* June 20, 2005.

Meek, James. "The Animal Research I Can't Defend." *Guardian,* May 2, 2002.

Nicolelis, Miguel A. L., and Mandayam A. Srinivasan. "Human-Machine Interaction: Potential Impact of Nanotechnology in the Design of Neuroprosthetic Devices Aimed at Restoring or Augmenting Human Performance." In *Converging Technologies for Improving Human Performance: Nanotechnology, Biotechnology, Informatio* Reynolds, Carson, and Rosalind Picard. "Ethical Evaluation of Displays that Adapt to Affect." http://affect.media.mit.edu/pdfs/04.reynolds-picard-ad.pdf (accessed December 23, 2004).

Schmorrow, Dylan D., and Amy A. Kruse. "DARPA's Augmented Cognition Program—Tomorrow's Human Computer Interaction from Vision to Reality: Building Cognitively Aware Computational Systems." Paper presented at the IEEE Seventh Conference on Human Factors and Power Plants, Scottsdale, AZ, September 2002.

Skagestad, Peter. "The Mind's Machines: The Turing Machine, the Memex, and the Personal Computer." *Semiotica* 111, no. 3/4 (1996): 217–43.

Zimmer, Carl. "Mind Over Machine." *Popular Science,* February 2004.

CHAPTER 3

Bloche, M. Gregg, and Jonathan H. Marks. "Doctors and Interrogators at Guantanamo Bay." *New England Journal of Medicine* 353, no. 1 (July 7, 2005): 6–8.

Central Intelligence Agency. "KUBARK Counterintelligence Interrogation," July 1963. http://www.gwu.edu/<tilde>nsarchiv/NSAEBB/NSAEBB122/CIA%20Kubark%201-60.pdf (accessed March 18, 2006).

Chase, Alston. *Harvard and the Unabomber: The Education of an American Terrorist.* New York: W. W. Norton, 2003.

CNN.com. "Humiliation Is Part of Interrogation," May 4, 2004. http://www.cnn.com/2004/US/05/03/ritz.cnna/ (accessed June 7, 2004).

Hersh, Seymour M. "Torture at Abu Ghraib." *New Yorker,* May 10, 2004.

Joint Non-Lethal Weapons Program. https://www.jnlwd.usmc.mil/mission/asp (accessed October 18, 2004).

Marks, John. *The Search for the "Manchurian Candidate": The CIA and Mind Control.* New York: Times Books, 1979.

Morgan, Charles A., III, Steve Southwick, Gary Hazlett, Ann Rasmusson, Gary

Hoyt, Zoran Zimolo, and Dennis Charney. "Relationships among Plasma De-hydroepiandrosterone Sulfate and Cortisol Levels, Symptoms of Dissociation, and Objective Performance in Humans Exposed to Acute Stress." *Archives of General Psychiatry* 61, no. 8 (2004): 819–25.

Smith, Theresa C. *No Asylum: State Psychiatric Repression in the Former USSR.* New York: New York University Press, 1996.

Zernike, Kate, and David Rohde. "Forced Nudity of Iraqi Prisoners Is Seen as a Pervasive Pattern, Not Isolated Incidents." *New York Times,* June 8, 2004.

CHAPTER 4

Churchland, Patricia Smith. *Neurophilosophy: Toward a Unified Science of the Mind-Brain.* Cambridge, MA: MIT Press, 1986.

Cunningham, William A., Marcia K. Johnson, Carol L. Raye, J. Chris Gatenby, John C. Gore, and Mahzarin R. Banaji. "Separable Neural Components in the Processing of Black and White Faces." *Psychological Science* 15, no. 12 (2004): 806–13.

Damasio, Antonio R. *Descartes' Error: Emotion, Reason, and the Human Brain.* New York: G. P. Putnam's Sons, 1994.

Dennett, Daniel C. *Consciousness Explained.* New York: Little, Brown, 1991.

Guchhait, R. B. "Biogenesis of 5-methoxy-N,N-dimethyltryptamine in Human Pineal Gland." *Journal of Neurochemistry* 26, no. 1 (1976): 187–90.

Schaffner, Kenneth F. "Neuroethics: Reductionism, Emergence, and Decision-Making Capacities." In *Neuroethics: Mapping the Field,* edited by Steven J. Marcus, pp. 27–33. Washington, DC: Dana Press, 2002.

Watson, Peter. *War on the Mind: The Military Uses and Abuses of Psychology.* New York: Basic Books, 1978.

Willingham, Daniel T., and Elizabeth Dunn. "What Neuroimaging and Brain Localization Can Do, Cannot Do, and Should Not Do for Social Psychology." *Journal of Personality and Social Psychology* 85, no. 4 (2003): 662–71.

CHAPTER 5

American Technology Corporation. "American Technology Corporation Awarded $1.088 Million Contract to Deliver Long Range Acoustic Devises (LRAD™) to U.S. Marine Corps Units," news release, February 26, 2004. http://www.atcsd.com/PressReleases/02_26_04.html (accessed January 26, 2005).

———. "American Technology Corporation Awarded $4.89 Million LRAD Order; ATC Supports U.S. Army with LRAD Deployments in Iraq," news release, December 15, 2004. http://www.tmcnet.com/usubmit/2004/Dec/1102214.htm (accessed January 25, 2005).

Defense Advanced Research Projects Agency. Fact File: A Compendium of DARPA Programs, August 2003. http://www.darpa.mil/body/news/2003/FINAL2003FactFilerev1.pdf (accessed January 26, 2005).

Engel, Adreas K., Christian K. E. Moll, Itzhak Fried, and George A. Ojemann. "Invasive Recordings from the Human Brain: Clinical Insights and Beyond." *Nature Reviews Neuroscience* 6, no. 1 (2005): 35–47.

Farah, Martha J., and Paul Root Wolpe. "Monitoring and Manipulating Brain Function: New Neuroscience Technologies and Their Ethical Implications." *Hastings Center Report* 34, no. 3 (2004): 35–45.

Federal News Service. Washington Daybook, Washington, DC, March 21, 2002 (NASA denial of mind reading goal).

Flam, Faye. "Your Brain May Soon Be Used Against You." *Philadelphia Inquirer,* October 29, 2002. http://www.prisonplanet.com/your_brain_may_soon_be_used_against_you.html (accessed January 21, 2006).

Gazzaniga, Michael. *The Ethical Brain.* New York: Dana Press, 2005.

Glimcher, Paul W. "Indeterminacy in Brain and Behavior." *Annual Review of Psychology* 56 (2005): 25–56.

Horgan, John. "The Myth of Mind Control." *Discover* 25, no. 10 (2004): 40–47.

Kirsch, Steve. "Identifying Terrorists before They Strike by Using Computerized Knowledge Assessment (CKA)." http://www.skirsch.com/politics/plane/ultimate.htm (accessed June 20, 2005).

Kozel, F. Andrew, Letty J. Revell, Jeffrey P. Lorberbaum, Ananda Shastri, Jon D. Elhai, Michael David Horner, Adam Smith, Ziad Nahas, Daryl E. Bohning, and Mark S. George. "A Pilot Study of Functional Magnetic Resonance Imaging Brain Correlates of Deception in Healthy Young Men." *Journal of Neuropsychiatry and Clinical Neurosciences* 16, no. 3 (Summer 2004): 295–305.

Langleben, D. D., L. Schroeder, J. A. Maldjian, R. C. Gur, S. McDonald, J. D. Ragland, C. P. O'Brien, and A. R. Childress. "Brain Activity during Simulated Deception: An Event-Related Functional Magnetic Resonance Study." *NeuroImage* 15, no. 3 (2002): 727–32.

Murray, Frank J. "NASA Plans to Read Terrorist's Minds at Airports." *Washington Times,* August 17, 2002.

National Institutes of Health. The Human Brain Project. http://www.nimh.nih.gov/neuroinformatics/index.cfm (accessed January 26, 2005).

Pasternak, Douglas. "John Norseen: Reading Your Mind—and Injecting Smart Thoughts." *U.S. News & World Report,* January 3, 2000.

Pearson, Helen. "Lure of Lie Detectors Spooks Ethicists." *Nature* 441:918–919.

Phillips, Helen. "Private Thoughts, Public Property." *New Scientist* 183, no. 2458 (July 31, 2004): 38–41.

Rilling, James K., David A. Gutman, Thorsten R. Zeh, Giuseppe Pagnoni, Gregory S. Berns, and Clinton D. Kilts. "A Neural Basis for Social Cooperation." *Neuron* 35, no. 2 (2002): 395–405.

Roskies, Adina. "Everyday Neuromorality." *Cerebrum* 6, no. 4 (2004): 58–65.

Sententia, Wrye. "Brain Fingerprinting: Databodies to Databrains." *Journal of Cognitive Liberties* 2, no. 3 (2001): 31–46.

Spence, Sean A., Mike D. Hunter, Tom F. D. Farrow, Russell D. Green, David H. Leung, Catherine J. Hughes, and Venkatasubramanian Ganesan. "A Cognitive Neurobiological Account of Deception: Evidence from Functional Neuroimaging." *Philosophical Transactions of the Royal Society B: Biological Sciences* 359, no. 1451 (2004): 1755–62.

Thompson, Sean Kevin. "The Legality of the Use of Psychiatric Neuroimaging in Intelligence Interrogation." *Cornell Law Review* 90, no. 6 (2005): 1601–37.

Wolpe, Paul Root, Kenneth R. Foster, and Daniel D. Langleben. "Emerging Neurotechnologies for Lie-Detection: Promises and Perils." *American Journal of Bioethics* 5, no. 2 (2005): 39–49.

CHAPTER 6

Baard, Eric. "The Guilt-Free Soldier." *Village Voice,* January 22–28, 2003.

Blackstone, Eric, Mike Morrison, and Mark B. Roth. "H_2S Induces a Suspended Animation–Like State in Mice." *Science* 308, no. 5721 (2005): 518.

Cortex Pharmaceuticals. "DARPA Extends Research Funding for the Prevention of Sleep Deprivation, Which Includes AMPAKINE™ Technology," news release, June 8, 2004. http://www.cortexpharm.com/html/news/04/06-08-04.html (accessed January 31, 2005).

DeRenzo, Evan G., and Richard Szafranski. "Fooling Mother Nature: An Ethical Analysis of and Recommendations for Oversight of Human-Performance Enhancements in the Armed Forces." *Airpower Journal* 11 (1997): 25–36.

Fukuyama, Francis. *Our Posthuman Future: Consequences of the Biotechnology Revolution.* New York: Picador, 2003.

Giles, Jim. "Alertness Drug Arouses Fears about 'Lifestyle' Misuse." *Nature* 436 (August 25, 2005): 1076.

———. "Electric Currents Boost Brain Power." *Nature.com,* October 26, 2004. http//www.nature.com/news/2004/041025/pf/041025-9_pf.html.

Graham-Rowe, Duncan. "World's First Brain Prosthesis Revealed." *New Scientist,* March 12, 2003. http://www.newscientist.com/article.ns?id=dn3488 (accessed January 25, 2005).

Groopman, Jerome. "Eyes Wide Open," December 3, 2001. http://www.jerome-groopman.com/eyes.html (accessed January 31, 2005).

Habeck, Christian, Brian C. Rakitin, James Moeller, Nicolaos Scarmeas, Eric Zarahn, Truman Brown, and Yaakov Stern. "An Event-Related fMRI Study of the Neurobehavioral Impact of Sleep Deprivation on Performance of a Delayed-Match-to-Sample Task." *Cognitive Brain Research* 18, no. 3 (2004): 306–21.

Hughes, James. *Citizen Cyborg: Why Democratic Societies Must Respond to the Redesigned Human of the Future.* New York: Westview Press, 2004.

Martin, Richard. "It's Wake-Up Time." *Wired* 11, no. 11 (November 2003). http://www.wired.com/wired/archive/11.11/sleep.html (accessed July 28, 2004).

Medical University of South Carolina (sleep deprivation studies). http://research.musc.edu/bp/centers_cair_bsl.html (accessed June 29, 2005).

Naam, Ramez. *More Than Human: Embracing the Promise of Biological Enhancement.* New York: Broadway Books, 2005.

Pitman, Roger K., Kathy M. Sanders, Randall M. Zusman, Anna P. Healy, Farah Cheema, Natasha B. Lasko, Larry Cahill, and Scott P. Orr. "Pilot Study of Secondary Prevention of Posttraumatic Stress Disorder with Propranolol." *Biological Psychiatry* 51, no. 2 (2002): 189–92.

President's Council on Bioethics. *Beyond Therapy: Biotechnology and the Pursuit of Happiness.* New York: Regan Books, 2003.

Schaffer, Amanda. "The Body Electric." *Slate,* January 11, 2005. http://www.slate.com/id/2112151 (accessed January 25, 2005).

Shachtman, Noah. "DARPA Offers No Food for Thought." *Wired News,* February 17, 2004. http://www.wired.com/news/medtech/0,1286,62297,00.html?tw=wn_tophead_1 (accessed January 28, 2005).

———. "Pentagon Revives Memory Project." *Wired News,* September 13, 2004. http://www.globalsecurity.org/org/news/2004/040913-pentagon-memory.htm (accessed February 22, 2005).

Shumyatsky, Gleb P., Gaël Malleret, Ryong-Moon Shin, Shuichi Takizawa, Keith Tully, Evgeny Tsvetkov, Stanislav S. Zakharenko, et al. "*stathmin,* a Gene Enriched in the Amygdala, Controls Both Learned and Innate Fear." *Cell* 123 (2005): 697–709.

Turse, Nick. "DARPA's Wild Kingdom." *Mother Jones,* March 8, 2004.

Walz, Chris. "Air Force Testing New Fatigue-Combating Drug." *Pentagram,* February 14, 2003. http://www.dcmilitary.com/army/pentagram/8_06/national_news/21626-1.html (accessed July 28, 2004).

Wessner, Charles W., ed. *Capitalizing on New Needs and New Opportunities: Government-Industry Partnerships in Biotechnology and Information Technologies.* Washington, DC: National Academies Press, 2001.

CHAPTER 7

American Technology Corporation. "American Technology Corporation Awarded $1.088 Million Contract to Deliver Long Range Acoustic Devices (LRAD) to U.S. Military Corps Units," news release, February 26, 2004. http://www.atcsd.com/PressReleases/02_26_04.html (accessed January 26, 2005).

———. "American Technology Corporation Awarded $4.89 Million LRAD Order; ATC Supports U.S. Army with LRAD Deployments in Iraq," news release,

December 15, 2004. http:///www.tncnet.com/usubmit/2004/Dec/1102214.htm (accessed January 25, 2005).

Bunker, Robert J. *Nonlethal Weapons: Terms and References.* INSS Occasional Paper 15. Institute for National Security Studies, U.S. Air Force Academy, Colorado, 1997. http://www.aquafoam.com/papers/Bunker.pdf (accessed June 24, 2005).

Cook, Joseph W., III, David P. Fiely, and Maura T. McGowan. "Nonlethal Weapons: Technologies, Legalities, and Potential Policies." *Airpower Journal,* special edition, 1995. http://www.airpower.maxwell.af.mil/airchronicles/apj/mcgowan.html (accessed March 3, 2005).

Coupland, Robin M. "Incapacitating Chemical Weapons: A Year after the Moscow Theatre Siege." *Lancet* 362, no. 9393 (October 25, 2003): 1346.

Defense-Aerospace.com, May 13, 2004. "American Technology Awarded $485,000 Contract to Deliver Long Range Acoustic Devices to U.S. Army." http://www.defense-aerospace.com/cgi-bin/client/modele.pl?prod=38802&session=dae.19818846.1144993152.RD81gMOa9dUAABtLIGU&modele=jdc_1 (accessed April 19, 2006)

GlobalSecurity.org. "Non-Lethal Weapons." http://www.globalsecurity.org/military/systems/munitions/non-lethal.htm (accessed June 21, 2005).

———. "Vehicle-Mounted Active Denial System (V-MADS)." http://www.globalsecurity.org/military/systems/ground/v-mads.htm (accessed June 21, 2005).

Grotius, Hugo. *The Law of War and Peace.* Bk. 2, chap. 1. 1949. Cited in Ziyad Motala and David T. ButleRitchie, "Self-Defense in International Law, the United Nations, and the Bosnian Conflict," *University of Pittsburgh Law Review* 57 (1995): 10 n. 75.

Jane's Information Group. "'Non-Lethal' Weapons May Have Significant Impact on International Law," news release, December 14, 2000. http://www.janes.com/press/pc001214.shtml (accessed June 29, 2005).

Joint Non-Lethal Weapons Program. https://www.jnlwp.com/mission.asp (accessed October 18, 2004).

Lakoski, Joan M., W. Bosseau Murray, and John M. Kenny. "The Advantages and Limitations of Calmatives for Use as a Non-Lethal Technique," Penn State College of Medicine Applied Research Lab, October 3, 2000. http://www.mindfully.org/Reform/Non-Lethal-Calmatives3octo0.htm (accessed March 4, 2005).

MacKay, H. Colin. "Non-lethal Weapons—Contributing to Psychological Effects in Operations Other Than War," Canadian Forces College. http://wps.cfc.dnd.ca/papers/amsc7/mackay.htm (accessed March 2, 2005).

National Research Council of the National Academies. *An Assessment of Non-Lethal Weapons Science and Technology.* Washington, DC: National Academies Press, 2003.

Sunshine Project. "Non-Lethal Weapons Research in the U.S.: Calmatives and Malodorants," July 2001. http://www.sunshine-project.de/infos/archiv/hintergrund/nr_08.pdf (accessed March 8, 2005).

———. "Pentagon Tests Ethnically-Targeted Crowd Control Weapons," news release, February 19, 2002. http://www.sunshine-project.org/publications/pr/pr190202.html (accessed March 8, 2005).

CHAPTER 8

Alibek, Ken, with Stephen Handelman. *Biohazard: The Chilling True Story of the Largest Covert Biological Weapons Program in the World—Told from Inside by the Man Who Ran It.* New York: Random House, 1999.

Atlas, Ronald. "Conduct of Biodefense Research." http://www.csis.org/tech/Biotech/events/050415_atlas.pdf (accessed May 17, 2005).

Harris, Elisa D., and John D. Steinbrunner. "Scientific Openness and National Security After 9/11." *CBW Convention Bulletin* 67 (March 2005).

King-Casas, Brooks, Damon Tomlin, Cedric Anen, Colin F. Camerer, Steven R. Quartz, and P. Read Montague. "Getting to Know You: Reputation and Trust in a Two-Person Economic Exchange." *Science 308,* no. 5718 (2005): 78–83.

Moodie, Michael L. *A Long-Term Response to Biological Terrorism. Issue Paper* 12. Center for the Study of the Presidency, Washington, DC, August 2005. http://www.thepresidency.org/pubs/IssuePaper12.pdf (accessed August 19, 2005).

National Research Council of the National Academies. *Biotechnology Research in an Age of Terrorism.* Washington, DC: National Academies Press, 2004.

National Science Advisory Board for Biosecurity. http://www.biosecurityboard.gov (accessed July 8, 2005).

Rosenberg, Barbara Hatch. "Defending Against Biodefence: The Need for Limits." *Disarmament Diplomacy* 69 (February–March 2003): 1–6. http://www.fas.org/bwc/papers/defending.pdf (accessed May 17, 2005).

INDEX

A

Abu Ghraib prisoner abuse scandal, 61–63, 64, 80
academia. *See* military-academic complex; *specific universities*
ACC (anterior cingulate cortex), 103, 104
acoustic devices, 146–151, 159
actions, intentional versus unintentional, 94–95, 103
active denial system (ADS), 153, 156–157
Adrian, Lord Edgar, 108
ADS (active denial system), 153, 156–157
Advanced Soldier Sensor Information System and Technology (ASSIST), 125–126
Advances in Military Medicine, 23
advertising-related mind control propaganda wars, 78–80
subliminal advertising, 146–147
AEC (Atomic Energy Commission), 25, 26–27
Afghanistan, 80, 144–145
African American soldiers, 65–66
aggression, 95–96
AI (artificial intelligence), 47–50, 54–55, 125–126
air traveler fMRIs, 76
Alibek, Ken (was Kanatjan Alibekov), 167–168
alienation focus in 1950s, 71–72
American Soldier, The (Stouffer), 65
American Technology Corporation (ATC), 147, 148
ampakines, 118
amphetamines, 114–115
amputees, problems of, 37–38

amygdala, 34–35, 93, 99, 128, 129–132
analog machine (Vannevar) versus digital machine (Turing), 49–50
anatomy of the brain, 34–36
animals. *See* experiments with animals
anterior cingulate cortex (ACC), 103, 104
anthrax, Pentagon and CIA supply of, 155
anthrax release in Soviet Union, 167–168
antimateriel non-lethal weapon technologies, 143
anti-nuclear weapons movement, 169–170
antipersonnel non-lethal weapon technologies, 143
antisleep drugs, 115–120
Applications of Biology to Defense Applications program, 12–13
Aquinas, St. Thomas, 160
Archimedes, 88
arms control efforts and NLWs, 159
artificial intelligence (AI), 47–50, 54–55, 125–126
Assessment of Non-Lethal Weapons Science and Technology (NAS), 141–142
Association of American Universities, 20
asymptomatic schizophrenia, 78
ATC (American Technology Corporation), 147, 148
Atlas, Ronald, 172–173
atomic bomb. *See* Manhattan Project
Atomic Energy Commission (AEC), 25, 26–27
AugCog (Augmented Cognition) program, 51–53
AugCog for Cockpit Design project, 51–52
AugCog International Conference, 181–182
Augustine, St., 160

195

BOOKS FOR GENERAL READERS

BRAIN and MIND

THE DANA GUIDE TO BRAIN HEALTH: *A Practical Family
Reference from Medical Experts* (with CD-ROM)

*Floyd E. Bloom, M.D., M. Flint Beal, M.D., and David J. Kupfer, M.D.,
Editors*

Foreword by William Safire

The only complete, authoritative family-friendly guide to the brain's development, health, and disorders. *The Dana Guide to Brain Health* offers
ready reference to our latest understanding of brain diseases as well as information to help you participate in your family's care.

16 full-color pages and more than 200 black-and-white illustrations.
Paper (with CD-ROM) 744 pp. 1-932594-10-8 • $25.00

THE CREATING BRAIN: *The Neuroscience of Genius*

Nancy C. Andreasen, M.D., Ph.D.

A noted psychiatrist and bestselling author explores how the brain achieves
creative breakthroughs, including questions such as how creative people
are different and the difference between genius and intelligence. She also
describes how to develop our creative capacity. 33 illustrations/photos.

Cloth 225 pp. 1-932594-07-8 • $23.95

THE ETHICAL BRAIN
Michael S. Gazzaniga, Ph.D.

Explores how the lessons of neuroscience help resolve today's ethical dilemmas, ranging from when life begins to free will and criminal responsibility. The author, a pioneer in cognitive neuroscience, is a member of the President's Council on Bioethics.

Cloth 225 pp.1-932594-01-9 • $25.00

A GOOD START IN LIFE: *Understanding Your Child's Brain and Behavior from Birth to Age 6*
Norbert Herschkowitz, M.D., and Elinore Chapman Herschkowitz

The authors show how brain development shapes a child's personality and behavior, discussing appropriate rule-setting, the child's moral sense, temperament, language, playing, aggression, impulse control, and empathy. 13 illustrations.

Cloth 283 pp. 0-309-07639-0 • $22.95

Paper (Updated with new material) 312 pp. 0-9723830-5-0 • $13.95

BACK FROM THE BRINK: *How Crises Spur Doctors to New Discoveries about the Brain*
Edward J. Sylvester

In two academic medical centers, Columbia's New York Presbyterian and Johns Hopkins Medical Institutions, a new breed of doctor, the neuro-intensivist, saves patients with life-threatening brain injuries. 16 illustrations/photos.

Cloth 296 pp. 0-9723830-4-2 • $25.00

THE BARD ON THE BRAIN: *Understanding the Mind Through the Art of Shakespeare and the Science of Brain Imaging*
Paul Matthews, M.D., and Jeffrey McQuain, Ph.D.
Foreword by Diane Ackerman

Explores the beauty and mystery of the human mind and the workings of the brain, following the path the Bard pointed out in 35 of the most famous speeches from his plays. 100 illustrations.

Cloth 248 pp. 0-9723830-2-6 • $35.00

STRIKING BACK AT STROKE: *A Doctor-Patient Journal*

Cleo Hutton and Louis R. Caplan, M.D.

A personal account with medical guidance from a leading neurologist for anyone enduring the changes that a stroke can bring to a life, a family, and a sense of self. 15 illustrations.

Cloth 240 pp. 0-9723830-1-8 • $27.00

UNDERSTANDING DEPRESSION: *What We Know and What You Can Do About It*

J. Raymond DePaulo Jr., M.D., and Leslie Alan Horvitz.

Foreword by Kay Redfield Jamison, Ph.D.

What depression is, who gets it and why, what happens in the brain, troubles that come with the illness, and the treatments that work.

Cloth 304 pp. 0-471-39552-8 • $24.95
Paper 296 pp. 0-471-43030-7 • $14.95

KEEP YOUR BRAIN YOUNG: *The Complete Guide to Physical and Emotional Health and Longevity*

Guy McKhann, M.D., and Marilyn Albert, Ph.D.

Every aspect of aging and the brain: changes in memory, nutrition, mood, sleep, and sex, as well as the later problems in alcohol use, vision, hearing, movement, and balance.

Cloth 304 pp. 0-471-40792-5 • $24.95
Paper 304 pp. 0-471-43028-5 • $15.95

THE END OF STRESS AS WE KNOW IT

Bruce McEwen, Ph.D., with Elizabeth Norton Lasley

Foreword by Robert Sapolsky

How brain and body work under stress and how it is possible to avoid its debilitating effects.

Cloth 239 pp. 0-309-07640-4 • $27.95
Paper 262 pp. 0-309-09121-7 • $19.95

IN SEARCH OF THE LOST CORD: *Solving the Mystery of Spinal Cord Regeneration*

Luba Vikhanski

The story of the scientists and science involved in the international scientific race to find ways to repair the damaged spinal cord and restore movement. 21 photos; 12 illustrations.

Cloth 269 pp. 0-309-07437-1 • $27.95

THE SECRET LIFE OF THE BRAIN

Richard Restak, M.D.

Foreword by David Grubin

Companion book to the PBS series of the same name, exploring recent discoveries about the brain from infancy through old age.

Cloth 201 pp. 0-309-07435-5 • $35.00

THE LONGEVITY STRATEGY: *How to Live to 100 Using the Brain-Body Connection*

David Mahoney and Richard Restak, M.D.

Foreword by William Safire

Advice on the brain and aging well.

Cloth 250 pp. 0-471-24867-3 • $22.95
Paper 272 pp. 0-471-32794-8 • $14.95

STATES OF MIND: *New Discoveries about How Our Brains Make Us Who We Are*

Roberta Conlan, Editor

Adapted from the Dana/Smithsonian Associates lecture series by eight of the country's top brain scientists, including the 2000 Nobel laureate in medicine, Eric Kandel.

Cloth 214 pp. 0-471-29963-4 • $24.95
Paper 224 pp. 0-471-39973-6 • $18.95

THE DANA FOUNDATION SERIES ON NEUROETHICS

HARD SCIENCE, HARD CHOICES: *Facts, Ethics, and Policies Guiding Brain Science Today*

Sandra Ackerman, Editor

Top scholars and scientists discuss new and complex medical and social ethics brought about by advances in neuroscience. Based on an invitational meeting co-sponsored by the Library of Congress, the National Institutes of Health, the Columbia University Center for Bioethics, and the Dana Foundation.

Paper 200 pp. 1-932594-02-7 • $12.95

NEUROSCIENCE AND THE LAW: *Brain, Mind, and the Scales of Justice*

Brent Garland, Editor. With commissioned papers by Michael S. Gazzaniga, Ph.D., and Megan S. Steven; Laurence R. Tancredi, M.D., J.D.; Henry T. Greely, J.D.; and Stephen J. Morse, J.D., Ph.D.

How discoveries in neuroscience influence criminal and civil justice, based on an invitational meeting of 26 top neuroscientists, legal scholars, attorneys, and state and federal judges convened by the Dana Foundation and the American Association for the Advancement of Science.

Paper 226 pp.1-932594-04-3 • $8.95

BEYOND THERAPY: *Biotechnology and the Pursuit of Happiness.*

A Report of the President's Council on Bioethics

Special Foreword by Leon R. Kass, M.D., Chairman.

Introduction by William Safire

Can biotechnology satisfy human desires for better children, superior performance, ageless bodies, and happy souls? This report says these possibilities present us with profound ethical challenges and choices. Includes dissenting commentary by scientist members of the Council.

Paper 376 pp. 1-932594-05-1 • $10.95

NEUROETHICS: *Mapping the Field. Conference Proceedings.*

Steven J. Marcus, Editor

Proceedings of the landmark 2002 conference organized by Stanford University and the University of California, San Francisco, at which more than 150 neuroscientists, bioethicists, psychiatrists and psychologists, philosophers, and professors of law and public policy debated the ethical implications of neuroscience research findings. 50 illustrations.

Paper 367 pp. 0-9723830-0-X • $10.95

IMMUNOLOGY

RESISTANCE: *The Human Struggle Against Infection*

Norbert Gualde, M.D., translated by Steven Rendall

Traces the histories of epidemics and the emergence or re-emergence of diseases, illustrating how new global strategies and research of the body's own weapons of immunity can work together to fight tomorrow's inevitable infectious outbreaks.

Cloth 260 pp. 1-932594-00-0 $25.00

FATAL SEQUENCE: *The Killer Within*

Kevin J. Tracey, M.D.

An easily understood account of the spiral of sepsis, a sometimes fatal crisis that most often affects patients fighting off nonfatal illnesses or injury. Tracey puts the scientific and medical story of sepsis in the context of his battle to save a burned baby, a sensitive telling of cutting-edge science.

Cloth 225 pp. 1-932594-06-X • $23.95
Paper 225 pp. 1-932594-09-4 • $12.95

ARTS EDUCATION

A WELL-TEMPERED MIND: *Using Music to Help Children Listen and Learn*

Peter Perret and Janet Fox

Foreword by Maya Angelou

Five musicians enter elementary school classrooms, helping children learn about music and contributing to both higher enthusiasm and improved academic performance. This charming story gives us a taste of things to come in one of the newest areas of brain research: the effect of music on the brain. 12 illustrations.

Cloth 225 pp. 1-932594-03-5 • $22.95
Paper 225 pp. 1-932594-08-6 • $12.00

FREE EDUCATIONAL BOOKS

(Information about ordering and downloadable PDFs are available at *www.dana.org.*)

PARTNERING ARTS EDUCATION: *A Working Model from ArtsConnection*

This publication describes how classroom teachers and artists learned to form partnerships as they built successful residencies in schools. *Partnering Arts Education* provides insight and concrete steps in the ArtsConnection model. 55 pp.

ACTS OF ACHIEVEMENT: *The Role of Performing Arts Centers in Education.*

Profiles of more than 60 programs, plus eight extended case studies, from urban and rural communities across the United States, illustrating different approaches to performing arts education programs in school settings. Black-and-white photos throughout. 164 pp.

PLANNING AN ARTS-CENTERED SCHOOL: *A Handbook*

A practical guide for those interested in creating, maintaining, or upgrading arts-centered schools. Includes curriculum and development, governance, funding, assessment, and community participation. Black-and-white photos throughout. 164 pp.

THE DANA SOURCEBOOK OF BRAIN SCIENCE: *Resources for Teachers and Students, Fourth Edition*

A basic introduction to brain science, its history, current understanding of the brain, new developments, and future directions. 16 color photos; 29 black-and-white photos; 26 black-and- white illustrations. 160 pp.

THE DANA SOURCEBOOK OF IMMUNOLOGY: *Resources for Secondary and Post-Secondary Teachers and Students*

An introduction to how the immune system protects us, what happens when it breaks down, the diseases that threaten it, and the unique relationship between the immune system and the brain. 5 color photos; 36 black-and-white photos; 11 black-and-white illustrations. 116 pp. ISSN: 1558-6758

PERIODICALS

Dana Press also offers several periodicals dealing with arts education, immunology, and brain science. These periodicals are available free to subscribers by mail. Please visit *www.dana.org.*